T0091808

ROUTLEDGE LIBRARY EDITIONS:
URBAN PLANNING

Volume 9

URBAN LAND & PROPERTY MARKETS IN GERMANY

URBAN LAND & PROPERTY MARKETS IN GERMANY

HARMUT DIETERICH, EGBERT DRANSFELD
AND WINRICH VOß

LONDON AND NEW YORK

First published in 1993 by UCL Press

This edition first published in 2018
by Routledge
2 Park Square, Milton Park, Abingdon, Oxon OX14 4RN

and by Routledge
711 Third Avenue, New York, NY 10017

Routledge is an imprint of the Taylor & Francis Group, an informa business

British Library Cataloguing in Publication Data
A catalogue record for this book is available from the British Library

ISBN: 978-1-138-49611-8 (Set)
ISBN: 978-1-351-02214-9 (Set) (ebk)
ISBN: 978-1-138-49475-6 (Volume 9) (hbk)
ISBN: 978-1-351-02574-4 (Volume 9) (ebk)

Publisher's Note
The publisher has gone to great lengths to ensure the quality of this reprint but
points out that some imperfections in the original copies may be apparent.

Disclaimer
The publisher has made every effort to trace copyright holders and would welcome
correspondence from those they have been unable to trace.

Urban land & property markets in Germany

Hartmut Dieterich
Egbert Dransfeld
Winrich Voß
Universität Dortmund

UCL PRESS

First published in 1993 by UCL Press

UCL Press Limited
University College London
Gower Street
London WC1E 6BT

The name of University College London (UCL) is a registered
trade mark used by UCL Press with the consent of the owner.

ISBN:
1-85728-049-0 HB

British Library Cataloguing-in-Publication Data
A catalogue record for this book
is available from the British Library.

Typeset in Times Roman.
Printed and bound by
Biddles Ltd, King's Lynn and Guildford, England.

CONTENTS

Preface xi
Abbreviations and acronyms xii

PART I Overview

1 Basic information 2
1.1 The constitutional and legal framework 2
1.2 The economic framework 8
1.3 The social framework 15
1.4 The land and property market and the building industry 20
1.5 Trends in spatial development 35

2 The policy environment 41
2.1 The period of reconstruction (1950s to early 1960s) 41
2.2 The period of enlargement of towns (1960s to mid-1970s) . . . 41
2.3 The period of city regeneration (mid-1970s to early 1980s) . . . 42
2.4 The period of inner urban development
    (Innenentwicklung) – 1980s 43
2.5 The current period (1988–) 44

3 The market situation in the new east German states 46
3.1 The special situation in the new Länder 47
3.2 The market 50

PART II The land market

4 The planning and legal framework 56
4.1 The legal environment 56
4.2 The financial environment 75
4.3 The tax and subsidy environment 85

5 The land-market process 105
5.1 Price-setting   105
5.2 Actors in the land market 108

6 The outcome of the urban land market 117
6.1 Demand for building land 117

6.2 Supply of building land 118
6.3 Transactions on the land market 119
6.4 Average sizes of building plots 122
6.5 Prices 123
6.6 Speculation in land and property 128

7 Case studies of the land market 129
7.1 Hildesheim 129
7.2 Stuttgart 149
7.3 Dortmund 173

PART III The urban property market
8 The framework of the urban property market . . . 190
8.1 The legal environment 190
8.2 Finance, tax and subsidies 193

9 The property-market process 196
9.1 Price-setting 196
9.2 The actors and their behaviour 196

10 The outcome of the urban property market 199
10.1 Demand and supply of property 199
10.2 Transactions in the property market 200
10.3 Prices 201

11 Case studies of the property market 204
11.1 Düsseldorf 204
11.2 Frankfurt 213
11.3 Köln 235

PART IV Evaluation
12 Evaluation of the functioning of the market for urban land
and property 252
12.1 Residential use 252
12.2 Commercial use 254

Bibliography 257
Legal instruments 266
Index of English terms 269
Index of German terms 273

PREFACE

This book aims to describe and explain the complicated system of operations of the German land and property market, focusing on both the framework and the functioning of the market. Component parts of the framework include the social and economic context, the legal framework, the administrative structures and the rules of taxation. Within the discussion of functioning, the actors and their current interests and the process of price-setting are decisive. The results generated by the market conditions are described as the outcome.

The system of the real-estate market was recorded according to a broad examination pattern adopted for the research programme, on which the structure of the book is based (see Foreword). It is not structured by market sectors (housing market, office market, etc.), but by specific component parts of the land and property market because the overall project, the "Functioning and Framework of the Urban Land and Property Markets in EC-Member States" aims at an evaluation and comparison of the market system with reference to the scope for intervention by public authorities. However, the differences between the market sectors are dealt with. The distinction between the land market and the property market was pursued in order to facilitate comparison between the different countries' systems. However, in Germany regulations are equally valid in both market sectors. In this report the situation of the market for urban land is the central interest. The property section is limited to additional points applicable only to that sector.

The bulk is devoted to presentation of information and its preparation and explanation, rather than evaluation. Judgements on German market conditions will be dealt with by direct comparison with other countries (Vol. 6 of this series). In the main, existing current information has been collected and presented, using the best available statistical data on the real estate market and its indicators in Germany. It is notable that the housing sector is considerably better served with data than the industrial or office sectors. The material as a whole presents an overview never previously published.

The six case studies contained in this book were selected on the basis of criteria such as market sectors, spatial issues, regional representation and typical constellations of actors. Their main purpose is to explain the functioning of the market. They are based on original empirical research, and set out in a self-contained form with contextual information on their respective regional situation, so that they may be read independently of the rest of the material.

All data was collected in the first part of 1991, so that most information is up to date to 1990. Specific conditions are changing fast because the current problematic market situation in Germany. However, this does not change the basic explanation of the German land and property market. This book is not, and cannot replace, a current market report.

A special problem is presented by the land and property market in the new eastern German *Länder*. Up to now a regular real-estate market hardly exists there. Moreover the data situation is far from comparable with that of the western *Länder*. Nevertheless since unification on 3 October 1990 the basic legal framework as described here, with the exception of some special rules, is also valid in the eastern part of Germany. In order to set out the specific legal arrangements and current market conditions at the time of writing, a separate chapter (Ch. 3) additional to the material from the original research report was written especially for this book.

HARTMUT DIETERICH   EGBERT DRANSFELD   WINRICH VOSS
DORTMUND, JULY 1992

ABBREVIATIONS AND ACRONYMS

BauNVO	Baunutzungsverordnung (land-use planning ordinance)
BBauG	Bundesbaugesetz (town planning law)
BfLR	Bundesforschungsanstalt für Landeskunde und Raumordnung (Federal Research Institute for Geography and Planning)
BGB	Bürgerliches Gesetzbuch (Civil Code)
BImSchG	Bundesimmissionsschutzgesetz (emmission control legislation)
BMBau	Bundesministerium für Raumordnung, Bauwesen und städtebau (Federal Ministry reponsible for spatial development, building and town planning)
BPlan	Bebauungsplan (building plan)
BRD	Bundesrepublik Deutschland (German Federal Republic)
CBD	central business district
CoC	Certificate of Completion
DDR	Deutsche Demokratische Republik (German Democratic Republic)
DM	Deutsche Mark
DSK	Deutsche Stadtentwicklungsgesellschaft (public development company)
EC	European Community
EHW	Einheitswert (standard value of property)
FNP	Flächennutzungsplan (Municipal preparatory land-use plan)
GA	Gütachterausschüsse für Grundstückswerte (Valuation Exchange)
GDP	gross domestic product
GFZ	floorspace ratio
GNP	gross national product
GRZ	plot ratio
ha	hectare
HbS	Hebesatz (property tax multiplier)
HOAI	legal scale of architects' charges
LEG	Landesentwicklungsgesellschaft (State development company)
m	metre
MWT	Mehrwertsteuer (VAT)
NH	Neue Heimat (housing association)
PMC	property management company
RDM	Ring Deutsche Makler (Association of estate agents)
SEM	Single European Market
StBauFG	Städtebauförderungsgesetz (urban renewal legislation)
SMZ	Steuermeßzahl (Federal property tax multiplier)
UVP	Umweltverträglichkeitsprüfung (environmental impact assessment)
UVPG	Umweltverträglichkeitsgesetz (environmental assessment law)
VAT	value added tax (MWT)
VDM	Verband Deutsche Makler (association of estate agents)
WHG	Wasserhaushaltsgesetz (water protection legislation)

PART I
Overview

CHAPTER 1

Basic information

1.1 The constitutional and legal framework

Constitution and organization of the state

Germany is a federal state organized as three autonomous tiers: the *Bund* (federal government), the *Länder* (state governments) and the *Städte und Gemeinden* (municipal governments). Separation of legislative, administrative and judicial powers is a basic principle. The constitutional principles on which the state is governed – in effect the constitution – are laid down in the *Grundgesetz* (basic law) promulgated on 23 May 1949, the latest changes to which provided for German reunification on 3 October 1990 (see Ch. 3).

Bund and Länder The Bund consists of 16 Länder, over which it has an umbrella-type function, and is itself vested with strong financial powers. The Länder are organized as independent states with separate constitutions, separate parliaments and a *Ministerpräsident* (minister-president or prime minister).

Three of the Länder are *Stadtstaaten* (city states): Berlin, Bremen and Hamburg. These states extend only a little way outside their respective cities and are completely surrounded by other Länder. Regional development of the Stadtstaaten is therefore dependent upon good co-operation with the neighbouring states. Co-operation is also necessary along the administrative borders of the larger Länder elsewhere, for example, in the Rhein–Main agglomeration, and in general it works well.

The legislative jurisdiction of the tiers is fixed in the basic law (Art. 70–75). In some fields only the Bund or the Länder are allowed to introduce legislation, while in others the federal parliament is empowered to delineate a legal framework within which the Länder operate, for example, in the case of allocation of land or regulation of spatial development (*Raumordnungsgesetz* 1989). Most areas of legal jurisdiction are subject to *konkurrierende Gesetzgebung*: both the Bund and the Länder have the jurisdiction to introduce legislation. Examples include legislation concerning the economy, land transactions, planning (*Baugesetzbuch* 1986), housing, traffic and the environment. However adoption of a law by the Bund limits the jurisdiction of the Länder. It is not uncommon for different pieces of legislation to complement each other and be enacted by different tiers.

There are three levels of the administrative hierarchy, of which the two lower are within the Länder structure. The upper level is represented respectively by the federal and state ministries. The Bund ministry responsible for the land and property market is the Bundesministerium für Raumordnung, Bauwesen und Städtebau (federal ministry for spatial development, building and town planning), normally abbreviated to BMBau; while, for example, the responsible ministries in the Land Nordrhein–Westfalen are the Ministerium für Stadtentwicklung und Verkehr (Ministry for Urban Development and Traffic) and the Ministerium für Bauen und Wohnen (Ministry for Building and Housing). The middle level in the larger Länder is the *Regierungspräsidenten*. This authority is responsible for all issues within a part of the state, known as the *Regierungsbezirk* (administrative district). Nordrhein–Westfalen, for example, is subdivided into five Regierungsbezirke. These prepare the plans for the development of the region, known as the *Gebietsentwicklungspläne* (see Ch. 4.1). Administration at the lower level within the Länder is the duty of the *Kreise* (counties), although these are constitutionally independent of the Länder level of government.

Kreise, Städte and Gemeinden   The third tier within the German federal system is the level of the municipalities, which in many respects is the most important as far as the land and property market is concerned. Each municipality has an elected council and all the rights and duties of local self-government (*Kommunale Selbstverwaltung*), protected by the basic law or constitution. The municipalities are responsible for the introduction of local laws (*Satzungen*), of which the *Bebauungsplan* (local plan) is an example central to the theme of this book, and are financially independent. Besides land-use planning, other responsibilities include the provision of schools and education, health and local transport policy (Hooper 1989: 264).

The constitution of the municipalities differs between northern and southern Germany as a result of the different policies of the Allied Occupying Powers after the Second World War. In the south, for example, one person is both mayor and chief of the local authority, while in the north these are separate offices.

The larger towns, known as *Kreisfreie Städte* (county-free towns), undertake all responsibilities. In addition, they carry out the lower-level state administrative duties. The other municipalities work together jointly within the counties (Kreise), which take responsibility for some tasks, such as waste disposal, public transport and cultural matters. In their function as the lower level of state administration, the counties produce landscape plans, are responsible for the cadastre, and have many duties concerned with environmental protection, including, for example, the *Gewerbeaufsicht* (factory inspectorate). The right to determine proposed land-uses, however, remains with each municipality.

The municipalities, together with the counties, are under the supervision of the Regierungspräsident or the administrative districts. In some respects this is important for the actors in the land and property market.

The constitution and organization of the federal state, the jurisdiction of the three tiers and the three separate powers are rather complicated and confusing. At all levels there are many peculiar features, intermediate authorities and sectoral jurisdictions too numerous to discuss in depth here. It is a system of widely spread participation and mutual control, but today it is very bureaucratic. In general, however, it has worked well.

Rights of ownership over land and property (tenure)

Definition of the ownership of a plot Ownership is an exclusive right. The owner of landed property can do with it whatever he wants and exclude all others from any influence on the property. The right of ownership is not unlimited, however (§903 Bürgerliches Gesetzbuch, BGB). The owner may do what he thinks fit only so long as the law or rights of third parties (for example, neighbours) do not stand against it. In theory, ownership of a plot of land includes the right to build on it. In practice, planning legislation limits the owner's right to construct buildings, demolish them or change the use of the land.

The owner may use the property himself, let it to tenants or sell it, and may even allow it to deteriorate. Alternatively it may be used as security for a loan.

The definition of ownership given at the beginning of the century in the BGB is still valid today, and was incorporated into Art. 14 of the Grundgesetz in 1949. The scope of the right of ownership has not changed, but greater stress is now placed upon the interests of the public. Restrictions have become tighter, particularly in relation to landscape protection and the conservation of historic buildings. Two of the most important restrictions for an owner, however, relate to planning law and rent law.

The right of ownership of a plot extends within the enclosed boundary of the Earth's surface and theoretically includes unlimited height above the plot, as well as below its surface to the centre of the Earth. The owner has, however, no rights over areas at a very great height or depth (§905 BGB). For example, mineral resources such as coal or iron ore are not part of the ownership, and groundwater is also excluded. Ownership rights include the extraction of building materials such as sand, and hunting and fishing rights.

Buildings on a plot are a component part of fixtures of the land and are owned by the same person. In German law there are no restrictions on property ownership by citizens of other countries. Moreover, no distinction is made between citizens of European Community member states and those of other countries. German law applies equally to owners of any nationality.

In addition to "complete" ownership or freehold as described above, there are other lesser rights over land, known as *Grundstücksgleiche Rechte*.

Hereditary long leasehold (Erbbaurechtsverordnung, 1919) Hereditary long leasehold is a substitute for ownership. An *Erbbaurecht* may be acquired if an individual cannot afford to pay the freehold price of a plot, or if the owner is unwilling to sell his property. An Erbbaurecht is the hereditary right to use the site and to have buildings on it. Buildings erected on a site with an Erbbaurecht are a component part or fixture of the Erbbaurecht, not of the land, and are owned by the person who bought the Erbbaurecht.

The duration of the hereditary right is usually between 80 years and 99 years for a residential building, and between 40 and 66 for industrial properties. During this time the person who buys an Erbbaurecht has, for all practical purposes, ownership rights. However, he has to pay a yearly rent to the owner. It is usual for the owner to arrange for a return from such land in the form of *Erbbauzins* (ground rent), which is typically about 4–5% for residential land, or more in the case of industrial land. In northern Germany it is also common to buy an Erbbaurecht with a single payment or premium.

The holder of the Erbbaurecht is permitted to sell it. Its value depends on the duration of the right until expiry and on the value of the buildings. As with the owner of a piece of land, the holder of the right also has to pay taxes. Municipalities, large estates and the Church most commonly establish and sell Erbbaurechte, while private persons and investors rarely use it. Erbbaurechte applies to only a small proportion of residential and industrial property.

Condominium law (Wohnungseigentumsgesetz, 1951) Ownership in the form of condominiums was created in 1951 and was intended to enable large sections of the population to become owners of their own flats. A building does not belong to any individual, but is split into separate flat ownerships (condominiums). This kind of ownership is created by a deed between the owners.

The *Wohnungseigentum* consists of two forms of ownership. Each individual flat itself belongs to the owner, who holds "flying freehold" ownership, while the common parts of the building are in joint freehold ownership. These common parts include roofs, stairwells, foundations and the plot itself. The owners of single dwellings are joint trustees in the *Eigentümergemein-*

schaft and are jointly responsible for the common parts of the property.

The Wohnungseigentum system is widespread and new buildings in cities are very often organized into this form of ownership. Often the flats are suitable for single people or couples without children. Old buildings that are modernized are also frequently converted into condominiums (see Köln case-study, Ch. 11.3).

Two further forms of property interests – tenancies and leasing contracts – are described in the next two sections. It should be noted, however, that they do not constitute forms of ownership with a legal interest in land, and therefore come under contract law rather than property-ownership law.

Tenancy (Miete or Pacht) A *Mietvertrag* or *Pachtvertrag* is a contract between an owner and an occupier. The rules for tenancies are found in the Civil Code (§535ff. BGB). If the contract is for a period longer than one year it must be in writing. In the housing sector such tenancies are very common, often more so than owner-occupation. The occupation of industrial property using Miete or Pacht tenancy contracts is at present uncommon, but their significance in this sector has grown considerably within the past two decades.

The tenancy contract may run for no more than 30 years, with the exception of special contracts that run for the lifetime of the tenant or of the landlord. In the industrial and office sectors it is common to agree on a fixed term (for example, 10 years) with a continuation clause containing an option to quit. This means that the contract will continue unless one of the parties applies to discontinue, giving notice to quit at a specified time in advance. If a tenancy is running for an indefinite time, as is usual in the housing sector, each party must give notice to quit three months before the end of a quarter (unless a longer time for notice has been agreed upon or is required by law).

The tenant has to pay rent, usually monthly. Rents can be freely negotiated in absence of rent control, especially for industrial property. Often the tenant is also required to pay a service charge for a share of maintenance costs and additional costs, such as public fees for the property, including street cleaning and other public service. It is possible to agree in advance on a rent increase every few years (i.e. a periodic rent review). No landlord is willing to bind himself to a certain rent for too long a time. In the housing sector, rent increases are limited in that they must be related to the average rent in the town. If there is a clause in the contract for commercial property allowing for an increase of the rent in line with the price of a specified commodity, an index of commercial rents or inflation as a whole, Federal Bank permission is necessary when drafting the contract.

The owner may sell the rented property to a third party. A sale does not affect the tenancy: the new owner has to abide by the terms of the tenancy (§571 BGB brings renting close to being a right in rem). The existing tenant must of course continue to pay the rent each month. A favourable tenancy may increase the investment value of a piece of property and result in a good sale price.

It is not common in Germany for a landlord to sell tenancies. The tenant is not entitled to sublet the rented property without the consent of the owner, but if the owner does not consent to this, the tenant may terminate the contract with three months' notice to the end of any quarter in the year.

For investors, such tenancy contracts offer the opportunity of a secure investment with a stable, gradual flow of income. Landlord may readily protect themselves from inflation by using rent review clauses, and the occupier is able to rent premises without having to purchase or construct them.

Many firms are reluctant to rent industrial property for financial reasons, since they cannot offer such property as security for mortgages. Conditions are, however, changing in this respect, and more large firms that are not in need of mortgages rent premises when opening new branches. German banks are not particularly willing to lend out money as venture capital, so new businesses often have to rent at first, as ownership is too burdensome at the start of operations. The traditional preference for ownership is to some degree a question of mentality, as businessmen prefer to operate from their own premises.

Leasing (Immobilienleasing) A special form of tenancy in the industrial and office sector, which is becoming more popular, is *Immobilienleasing*. In Germany the term "leasing" is not used in the context of a simple Mietvertrag or Pachtvertrag. The owner, who often buys property for this purpose alone, lets property for a fixed sum consisting of the price of the property as well as all costs, interest payments, etc., which is payable in instalments. The occupier is fully responsible for the property. There is usually a fixed period for the leasing arrangement, often with an option to extend the leasing contract or to buy the property. It is not uncommon for a company to sell property so it can lease it back after conversion in good condition, thus helping cash flow and the strength of the balance sheet. Leasing is not only a kind of tenure, but some businesses provide an Immobilienleasing service, for example in the case of serviced offices. Immobilienleasing is usual only for new industrial or office development or for totally refurbished buildings.

Inheritance customs When considering ownership rights, the question of inheritance is also of interest. Two different types are used, particularly in the agricultural sector. In the north of Germany the *Anerbenrecht* dominates. This is the system whereby one heir inherits the land and property, and has

7

to buy out the other heirs. A more common system in the south of Germany is the *Realteilung*, whereby heirs inherit equal proportions of a parcel of land and property. This results in very small plots (see Stuttgart case study, Ch. 7.2).

The legal rules in the case of intestate succession define three degrees of claim:

○ first claim: children, grandchildren and great-grandchildren;
○ second claim: parents and their children (sisters and brothers);
○ third claim: grandparents and their children.

A marriage partner has a separate claim that comes to a quarter of the inheritance within the first claim and to one half within the second and third claims. If a marriage contract exists (the legal rule is equal division of property acquired after marriage), the claim of the marriage partner may be higher or lower. If there is a will, the heirs of first claim together with the parents and the marriage partner all have a claim of one half of the above-mentioned shares (or the legitimate portion).

The testator may furthermore determine the first heir for life, with reversion to a following heir on the death of the first heir (*Nacherbe*). So a property can be bound to a family and may not be sold. Where more than one heir inherits a property, they are described as joint heirs (*Erbengemeinschaft*). The heirs only have joint sale and management rights over the property at their disposal, in other words each must agree to any dispositions. This form of ownership is very inflexible and is often a hindrance in the land and property market.

1.2 The economic framework

This section contains some basic economic indicators and data on economic development that give an indication of the current economic situation in Germany and of changes in the past decade. Most of the data do not directly influence the land and property market, but are indicators of the general economic climate in which it functions. Furthermore, the data will also enable rough comparisons to be made between the base positions of the German economy and those of other countries in the Single European Market.

Table 1.1 shows changes in Gross National Product (GNP) in terms of market prices and real prices between 1980 and 1990. The Gross Domestic Product (GDP) is approximately 1% smaller than the GNP. There has been a steady increase in productivity since the mid-1980s, as Table 1.2 indicates. Figure 1.1 shows the long-term development of the inflation rate using the cost-of-living index. Note that in 1991 there was a greater increase in infla-

Table 1.1 Changes in GNP.

Year	GNP in market prices DM billion	GNP in market prices % increase	GNP in real prices 1980 DM billion	GNP in real prices 1980 % increase
1980	1,485.2	6.3	1,485.2	1.5
81	1,545.1	4.0	1,485.8	(0.0)
82	1,597.1	3.4	1,471.0	(1.0)
83	1,680.4	5.2	1,498.9	1.9
84	1,769.9	5.3	1,548.1	3.3
1985	1,844.3	4.2	1,578.1	1.9
86	1,945.2	5.5	1,614.7	2.3
87	2,015.6	3.6	1,639.8	1.6
88	2,123.2	5.3	1,700.5	3.7
89	2,261.3	6.5	1,766.0	3.9
1990	2,447.7	8.2	1,847.6	4.6

Source: Wirtschaft und Statistik 4/1991: p.229

Table 1.2 Growth of productivity.

Year	GDP per employee in real prices (1980) DM/employee	GDP per employee in real prices (1980) % change
1984	58,418	
1985	59,137	1.2
1986	59,697	0.9
1987	60,265	1.0
1988	62,004	2.9
1989	63,168	1.9
1990	64,360	1.9

Source: Wirtschaf und Statistik 3/1990:158;
1/1991:18

tion than in previous years. The cumulative inflation rate over the five-year period 1986–90 was 6.7%, i.e. average yearly inflation of only 1.34%. This contrasts with cumulative inflation over the 10-year period 1981–90 of 29%, reflecting the relatively high inflation rate in the early 1980s. Detailed information about one factor influencing inflation – interest rates – is given in Chapter 4.2.

Figure 1.2 shows changes in the number of people available for work between 1980 and 1990, which overall increased by 9%. The other graph

9

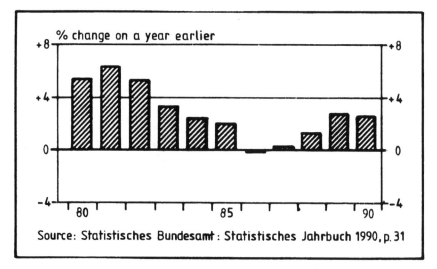

Source: Statistisches Bundesamt: Statistisches Jahrbuch 1990, p. 31

Figure 1.1 Cost-of-living index – all private households.

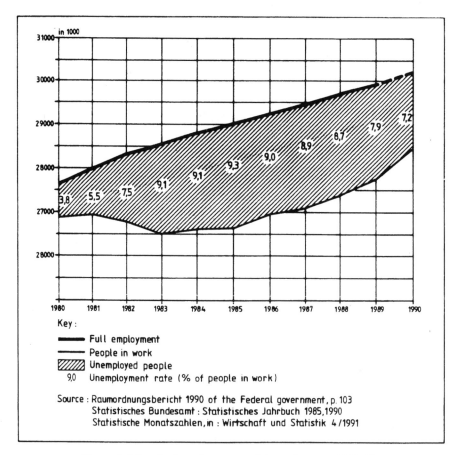

Key:
— Full employment
— People in work
▨ Unemployed people
9,0 Unemployment rate (% of people in work)

Source: Raumordnungsbericht 1990 of the Federal government, p. 103
Statistisches Bundesamt: Statistisches Jahrbuch 1985,1990
Statistische Monatszahlen, in: Wirtschaft und Statistik 4/1991

Figure 1.2 Level of employment and unemployment 1980–90.

illustrates the number of people actually in work. After a decrease in the mid-1980s, this has risen strongly in the past two years and the trend continues today. The difference fairly accurately represents the unemployment rate.

The relative importance of different parts of the economy are shown in Table 1.3 by percentage of *Bruttowertschöpfung* (gross wealth creation) and of employees and in Table 1.4 by rate of investment. The economy is divided for this purpose into three sectors: (I) agriculture/forestry/fishing, (II) industry (*warenproduzierendes Gewerbe*) and (III) trade and services (*Dienstleistungen*) Bruttowertschöpfung is a measure about 7% smaller than the GDP, and does not include customs duty and some value-added tax (VAT).

Table 1.3 Importance of the economic sectors.

Year	% of Bruttowertschöpfung			% of employees		
	I	II	III	I	II	III
1980	2.2	46.2	51.6	5.3	45.3	49.4
1982	2.5	44.6	52.9	5.4	43.2	52.7
1984	2.1	43.8	54.1	5.2	41.8	53.0
1986	1.9	43.9	54.2	4.6	41.1	54.3
1988	1.7	42.4	55.9	4.2	41.1	54.7
1990	1.7	42.5	55.8			

Source: Statistisches Jahrbuch 1990:30,569f; Wirtschaft und Statistik 1/1991:19

Table 1.4 The rate of investment in the economic sectors.

Year	% of GDP	% of economic sector			
		I	II	III	state
1980	23.3	2.7	25.0	57.4	14.9
1982	19.3	2.7	25.3	58.5	13.5
1984	20.2	2.6	23.9	61.8	11.7
1986	19.4	2.5	27.0	58.8	11.7
1988	19.9	2.3	86.4		11.3
1990	22.1				10.9

Source: Statistisches Jahrbuch 1990:576f; Wirtschaft und Statistik 4/1991:234; 1/1991:22

Table 1.4 shows the investment shares of the government as well as those of the three economic sectors; Table 1.5 shows the composition of building investment (based on market prices). Building investment is divided into equipment and investments in buildings.

Table 1.5 Components of the rate of investment.

Year	equipment	% of the investments			State
		total	buildings resident	other	buildings
1980	37.9	62.1	29.9	32.1	14.2
1982	38.3	61.6	29.8	31.7	14.3
1984	38.7	61.2	31.7	29.5	11.7
1986	42.8	57.2	27.3	29.9	10.7
1988	44.1	55.9	26.4	29.5	10.2
1990	45.8	54.2	30	25	9.2

Source: Statistisches Jahrbuch 1990:576f; Wirtschaft und Statistik 4/1991:234; 1/1991:22

In 1990, federal government revenue was DM1,064 billion. This included DM572 billion in taxes and DM410 billion in social security contributions. Expenditure was DM1,115 billion, including DM547 billion in transfers, DM447 billion for the state's consumption and DM56 billion in investments. This left a deficit corresponding to 2% of GDP (DM51 billion) in 1990 (Strohm 1991: 26). There has been a deficit in each year since the mid-1970s, apart from 1989.

Changes in wages and salaries, which determine total household income, are an important indicator of demand for property. Table 1.6 shows the changes in monthly wages and salaries between 1980 and 1990.

Table 1.6 Monthly wages and salaries.

Year	% increase of wages on a year earlier		
	in market prices	in real prices 1980	
		workers	employees
1980		0.0	1.7
1981		(1.9)	(1.3)
1982	4.2	(1.6)	(0.3)
1983	3.4	(0.6)	0.0
1984	2.3	1.0	0.9
1985	3.2	1.2	1.7
1986	3.2	3.3	3.6
1987	3.8	3.1	3.5
1988	2.6	2.9	2.4
1989	2.5	0.8	0.5
1990	3.6	1.4	1.4

Source: Wirtschaft und Statistik 4/1991:285–291

Table 1.7 gives the average social security contributions and taxes on monthly income.

Table 1.7 Components of the gross income from employment.

	Components	1990	%	1985	%
	Gross income per month	4,136	100	3,505	100
minus	Social security contributions of the employer	795	19	681	19
equals	Gross wage/ salary	3,340	81	2,824	81
minus	Income tax and social security contributions of the worker/employee	565	14	513	15
		499	12	415	12
equals	Net wage/salary	2,276	55	1,896	54

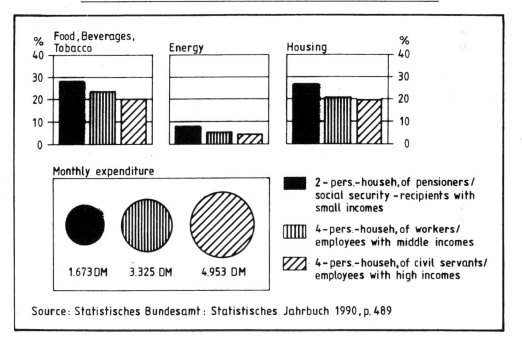

Figure 1.3 Aspects of the total expenditure for private consumption in 1989.

Weekly working hours are not standardized in Germany and they depend on the sector and on collective agreements between employers and employees in a particular sector. On average in 1990 an employee worked 38.17 hours per week as compared with 39.6 in 1985. The current figure varies between 39.9 hours in the mining industry and 37 hours in the capital goods industry (Heinlein 1991: 287). In some sectors a 35-hour week has been agreed, to start in 1995. Thus it would appear that leisure time will continue to increase.

Annual working days are determined by the number of public holidays, annual leave and average days lost due to illness. There are approximately 10 public holidays in Germany, the exact number depending on local religious and state practices. Annual leave is covered by collective agreements and is generally 29 days for an employee aged 30.

Table 1.8 Private savings in the FRG, 1980–89

Year	Billion DM	% of available income
1980	137	14
81	154	
82	147	
83	134	12.2
84	147	
85	151	
86	167	13.5
87	176	
88	186	
1989	190	13.6

Source: Deutsche Bundesbank; Informationsdienst des Volksheimstattenwerkes 10/1990

Table 1.9 Wealth in savings in the FRG.

	Breakdown 1989 Bn DM	Breakdown 1990 Bn DM
bank deposit accounts	694	680.4
life assurance policies	592	633.8
fixed-interest securities	440	515.6
savings certificates & short-term fixed deposits	330	387.7
company pension funds	217	228.8
cash and current accounts	213	229.4
shares	185	182.2
building society deposits (Bausparen)	121	168.5
other Investments	13	24.4
total	2,805	3,050.8

Source: Westfalische Rundschau vom 29/6/1990 und 4/6/1991.

In 1989, strikes or lockouts involved 44,000 employees (0.2% of the total workforce). Most strikes lasted less than a week. The average in the years 1983–8 totalled 45 days per 1,000 employees.

The rate of savings also affects demand for land and property. On average, the rate is 13–14% of the net annual income of a private household. Tables 1.8 and 1.9 describe this in more detail.

1.3 The social framework

Demographic development

Population At the end of 1989 the combined populations of the then German Democratic Republic (16.4 million) and Federal Republic (62.7 million) totalled 79.1 million. On average, the population in the former Federal Republic (BRD), or old Länder, consisted of 92% Germans and 8% ethnic minorities.

The following figures apply only to the old Länder. The population decreased at the beginning of the 1980s, and since 1985 has slowly increased. There was an increase of half a million (0.8%) in 1988 and 1 million (1.6%) in 1989 caused by immigration from eastern Europe and the former German Democratic Republic (DDR). These were the largest changes since the early 1970s. In 1989 the population density was 248 inhabitants per square kilometre. Regional variations in population density and other demographic data are shown in Table 1.10.

Figure 1.4 shows age distribution in Germany and forecast trends. The number of retired people will increase, while the number in work will decline, as will the number of young people. It is expected, however, that the new immigrants from eastern Europe will alter these trends.

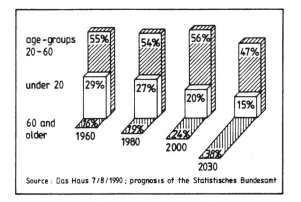

Figure 1.4 Age distribution.

Table 1.10 Area, population, employees and population density (1988).

Type of region	Area 1988 sq km	Area 1988 % of FRG	Population 1988 in 1000	Population 1988 % of FRG	Employees paying social security contributions 1988 in 1000	Employees paying social security contributions 1988 % of FRG	Density of population 1988 people per sq km	Density of settlement 1988 people per sq km, built-up and traffic area	Built-up and traffic area 1988 % of total area
Regions with large urban agglomerations	67,563	27.2	34,396	55.7	12,544	59.0	509	2,758	18.5
within these, old industrialized regions	8,714	3.5	6,764	11.0	2,079	9.8	776	2,968	26.2
Regions with major urban centres	96,253	38.7	17,611	28.5	5,671	26.7	183	1,622	11.3
Rural regions	84,862	34.1	9,709	15.7	3,050	14.3	114	1,365	8.4
FRG	248,678	100.0	61,715	100.0	21,265	100.0	248	2,027	12.2

Source: Laufende Raumbeobachtung der Bundesforschungsanstalt fur Landeskunde und Raumordnung.

The increase in population at the end of the 1980s was the result of immigration rather than natural increase of the existing population. The migration rate beyond the borders of Germany (*Außenwanderung*) is related to economic development and, while in the early 1980s there was a negative migration rate, it has grown heavily since 1985 as a result of changes in central and eastern Europe. In 1989, 35% of the immigrants came from the DDR, especially in November after the Berlin Wall was opened up. Ethnic Germans (*Aussiedler*) from other eastern European countries made up 40%, and 25% came from other countries. The government initially attempted to distribute immigrants proportionally to the Länder, but immigrants are now more free to settle where they choose and most live in the conurbations, particularly inner-city areas.

Ethnic minorities Overall, the ethnic minority population of western Germany increased steadily in the 1980s, except for a decline in the years after the economic recession of 1983–4. The distribution by Länder is more variable. In 1989 more than 10% of the population in the Stadtstaaten (Berlin, Bremen, Hamburg), Baden-Württemberg and Hessen were ethnic minorities, whereas in the northern Länder, such as Rheinland-Pfalz and Saarland, they made up less than 5%. In the old Länder in the western part of Germany minorities averaged 7.8% of the population (Fleischer 1990: 542).

More important is the distribution between urban and rural regions. In 1987, 50% of the ethnic minority population lived in towns within the metropolitan areas: 24.4% in Frankfurt, 18.5% in Stuttgart and 18.1% in München. They often live in old and unmodernized flats with low rents and are concentrated in certain districts of the urban area. The number of inhabitants in the inner cities is no longer declining because of this influx (Böltken & Schön 1989: 835).

Structure of households In 1989 there were 27.8 million private households in the BRD. This number has increased faster than the population in recent years, averaging 1.4% annual growth. The persons per household has decreased from 3.0 in 1950 to 2.74 in 1970 and to 2.24 in 1989. On average, households in cities with more than 100,000 inhabitants had less than two persons in 1989. The smallest households are in München and Berlin, and the largest, with more than three persons, are in the northwest and in rural areas of Bayern.

The number of single-person households has increased rapidly. Figure 1.5 shows the importance of this group, which makes up more than one-third of all households. Of these, 40% consist of people over the age of 65, mostly women with low disposable incomes. Nearly 30% are in the 25–45 age range. This group is dominated by men, and numbers have increased in the past decade. Their disposable income is high, 20% having a monthly net income of more than DM2,500. And 10% of people in single households are under the age of 25.

A quarter of single households are situated in the larger cities of more than 500,000 inhabitants. One-third of these fall into the 25–45 age group, and nearly half the households in these cities are single-person. Another quarter

% on total households:

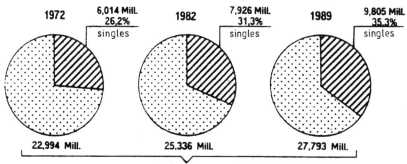

private households

Source: Wirtschaft und Statistik 10/1990, p. 703

Figure 1.5 Singles 1972, 1982 and 1989.

of the single-person households are in towns of between 20,000 and 100,000 inhabitants, where they constitute one-third of all households. Even in small municipalities with fewer than 5,000 people, a quarter of all households are single-person. All the figures given are for 1989 (Pöschl 1990: 703–8).

Mobility

Another important factor for the land and property market is population movement within Germany (*Binnenwanderung*). The movements between the old Länder totalled almost 800,000 people in 1989. Because of the north/south divide most people moved to the south German Länder of Baden-Württemberg and Bayern (30,000 to each), to Hessen (15,000) and to Rheinland-Pfalz (10,000). The other Länder gained small numbers of new inhabitants, except for Niedersachsen which lost 108,000 inhabitants (1.5% of its population) to other Länder. Niedersachsen also gained 214,000 immigrants and so overall the population increased (Sommer & Fleischer 1991: 86). In recent years there has been a marginal net migration from rural areas to urban regions. It is only in the under-25 age group that there is still a movement into urban areas (BMBau 1990a: 61).

Movements over short distances are generally for housing and family reasons, while longer-distance moves are more often due to changes in workplace.

In addition to the 800,000 people moving between the Länder and the 1 million surplus of immigrants and emigrants in 1989, a further 2.1 million people moved within a Land from one municipality to another (Sommer & Fleischer 1991: 86). The number of movements within a municipality is expected to be even higher. Surveys carried out as part of the *Innerstädtische Raumbeobachtung* project of the Bundesforschungsanstalt für Landeskunde und Raumordnung (BfLR), which included 27 cities, revealed that there were approximately 10 people moving at any one time per 1,000 inhabitants. This extrapolates to a figure of 600,000 people moving within Germany's municipalities (Böltken & Schön 1989: 837). This number is the sum of the balances within six age-groups. The number of single moves is higher.

Mobility varies according to the economic situation, the age-group concerned and whether Germans or ethnic minorities are involved. In general, however, mobility has decreased since the 1970s. Movements over large distances have declined to 60% of those in the early 1970s, and over medium distances have declined by one-third. Movements over short distances have increased since the late 1970s. Today 50% of moves are within 50 km, with another 20% between 50 and 100 km (BMBau 1990a: 64).

There has been a further decrease in mobility following a recent slump in the housing market. Between 1988 and 1990 only 2.7 million households (about 4% of all households per year) moved. This figure varies with the age of the household: whereas more than 12% of young households under the age of 35 changed their dwellings each year, only 1% of households with persons over 60 did so (GEWOS 1990a).

Changes in social values

In any society social values change over time. For example, in the years after the Second World War, material considerations were of the highest priority in Germany, while today such things are taken for granted. Young people strive for travel, more leisure time and the opportunity to realize individual goals. It should be noted that former priorities – security of employment and income, education and personal security – are still important, but new demands now compete.

Changing social values that directly influence the land and property market include (BMBau 1990a: 14):

○ After an acceptable quality of work premises, the quality of, and provision for, leisure activities in a town is becoming more important for the locational decisions of an employee.

○ The behavioural rôles of women have changed because of the feminist movement. Most young women aim to work, and many already do. Many opt to live alone. If two people in a household are working, especially if they have a family, the organization of time as well as access to services such as schools, playgroups or work places is more important. The mobility of households has decreased, whereas income is increasing.

○ Young people leave their parental homes earlier, whether or not they receive professional training or are in employment.

○ In general the importance of privacy and individual freedom is increasing. The expectations over one's flat, the cultural scene or one's social contacts are growing in importance. This also decreases mobility over large distances.

○ Environmental issues are another aspect of change in German society. Clean air, soil and food are becoming more important in the behavioural patterns and the expectations of individuals, but demands on the state are also growing. For example, restrictions on car usage have not yet been accepted but are being discussed. A survey based on studies in 1987 showed that clean air was considered the most important factor influencing wellbeing (65%, followed by medical care, 55%) (BMBau 1990a: 15).

1.4 The land and property market and the building industry

Owner-occupation

In western Germany the proportion of flats in owner-occupation increased between 1968 and 1987 from 34.3% to 37.5% (BMBau 1990a: 91). In the unified Germany the proportion currently in owner-occupation may be assumed to have decreased since a large part of the housing stock in the five new states is owned by public authorities or housing associations. In 1988 the proportion of non-commercial property in ownership was nearly 48%. This figure includes owners who do not live in their own houses or apartments (approximately 9%) and owners of land that has not been built upon. In 1988 11.5 million private households out of 24.7 million possessed *Grundvermögen* (real property: i.e. owner-occupied houses, apartments, building land) (Euler 1991: 278). This shows that most Germans remain tenants.

Figures on ownership will be more meaningful if the spatial regions, called *Raumordnungsregionen*, are distinguished. In inner-city areas the level of ownership is no more than 18% and is clearly below the federal average. The main reason is the high rate of one- or two-person households, which predominantly occupy rented flats and often do not seek ownership. By contrast, the rate of ownership is above the national average in the countryside surrounding large urban areas and in regions with major urban centres. In the more densely populated rural areas, owner-occupation accounts for 43.2% of all households and in the less-populated areas the figure is between 50% and 52%. Peripheral regions also show high levels of owner-occupation of approximately 51% (BMBau 1990a: 90).

Across the country there is much variation in the level of owner-occupation. The Stadtstaaten (city states) of Berlin and Hamburg have a very low level of ownership, while the states of Saarland and Baden-Württemberg have the highest levels: 65% and 60%, respectively. In the southern states the rural areas surrounding cities have the highest levels of owner-occupation, while in the northern states the peripheral regions show the highest levels.

A meaningful evaluation of ownership rates is possible only if they are compared with the distribution of ownership. Table 1.11 differentiates between several types of property, and the structure of households, their incomes and ages.

Among all types of property the one- and two-family houses are predominant. Approximately 82% of all households with real property possess this type, while only 6% have houses occupied by several families and 16.7% possess apartments. There is, furthermore, a clear correlation between the share of owners and the number of people per household. The rate of ownership is increasing among the larger households. For example, only a quarter of one-person households, compared with three-quarters of households with more than five persons, have real properties. Families with more than three children are urged to be property owners because it is difficult for them to find suitable flats to rent.

In terms of social position, the distribution of ownership shows the expected trends: 92% of farm households and 73.5% of all households of self-employed persons are property owners while the figure for unemployed persons is 22%.

Another aspect that has to be taken into account is the number of households on the move and the form of ownership they are striving for. Currently only 5% of moving households are owner-occupiers, while there are more than 18% after moving (GEWOS 1990b: 8).

To become an owner-occupier, it is usually necessary to save for some time. Thus only 5.4% of households with an average age of head of household of up to 25 years and 28% of households with heads of between 25 and 35 years possess their own property (Euler 1991: 279). The average age of people who acquire newly built properties is approximately 39, while it is nearly 41 years for older properties. The average age of people who bought property has decreased since 1985 by two years (Raschke 1991: 35) and the average number of wage earners is now up to 1.5 persons per household.

Table 1.11 Private households with real property.[1]

	total households	households with real property 1000's	%	one- and two-family house 1000's	%	owner of house for several families 1000's	%	condominiums 1000's	%
total	24,684	11,529	46.7	9,466	82.1	691	6.0	1,925	16.7
			with regard to size and type of households						
1 person	8,463	2,043	24.6	1,413	67.8	167	8.0	578	27.7
2 persons	7,810	4,000	51.2	3,298	82.5	236	5.9	652	16.3
3 persons	4,049	2,399	59.3	2,026	84.5	129	5.4	369	15.4
4 persons	3,138	2,112	67.3	1,856	87.9	116	5.5	262	12.4
5+ persons	1,224	936	76.4	873	93.3	42	4.5	64	6.8
			with regard to monthly net income						
DM									
below 1200	3,654	749	20.5	613	81.8	37	4.9	91	12.1
1200–1600	2,870	769	26.8	621	80.8	54	7.0	104	13.5
1600–2000	3,294	1,076	32.7	859	79.8	48	4.5	190	17.7
2000–2500	3,842	1,687	43.9	1,374	81.4	86	5.1	245	14.5
2500–3000	2,929	1,549	52.9	1,255	81.0	78	5.0	244	15.8
3000–4000	3,990	2,442	61.2	1,982	81.2	127	5.2	399	16.3
4000–5000	1,913	1,391	72.7	1,149	82.6	86	6.2	261	18.8
5000–7000	1,411	1,171	83.0	979	83.6	93	7.9	270	23.1
7000–10000	342	287	83.9	245	85.4	42	14.6	82	28.6
10000–25000	99	93	94.2	81	87.1			31	33.3
			with regard to social position						
farmer	328	302	92.1	295	97.7	15	5.0		
self-employed	1,371	1,007	73.5	815	80.9	163	16.2	219	21.7
civil servant	1,599	908	56.8	731	80.5	35	3.8	182	20.0
white-collar worker	5,514	2,664	48.3	2,025	76.0	121	4.5	657	24.7
blue-collar worker	4,856	2,354	48.5	2,028	86.2	80	3.4	269	11.4
unemployed	964	212	22.0	175	82.5			32	15.1
not employed	10,053	4,081	40.6	3,398	83.3	265	6.5	559	13.7
			with regard to age						
age									
below 25	808	44	5.4	26	59.1				
25-35	3,935	1,111	28.2	820	73.8	43	3.9	223	20.1
35-45	4,122	2,259	54.8	1,844	81.6	91	4.0	413	18.3
45-55	4,717	2,840	60.2	2,356	83.0	196	6.9	523	18.4
55-65	4,224	2,435	57.6	2,077	85.3	160	6.6	349	14.3
65-70	2,108	1,108	52.6	944	85.2	71	6.4	149	13.4
70+	4,770	1,733	36.3	1,400	80.8	128	7.4	259	14.9

Source: Wirtschaft und Statistik 4/91:278.

Note: 1/ Excluding foreign households and those with monthly income over DM 25,000.

Ownership differs between new, used and inherited buildings. New one- or two-family houses are purchased by 59% of households, while only 20% of households purchase second-hand houses and 21% inherit properties. By contrast, multi-family houses predominate in the second-hand market. Approximately 40% of households inherit such buildings, 28% buy used multi-family houses and 33% purchase new houses. Apartments have become more important in the second-hand market. Currently about 40% of all households buy used apartments, while 53% buy new ones. Inherited apartments are of less importance because this kind of property has been in existence only since 1951.

Inherited properties, and therefore the second-hand market, will become more important as properties built since the Second World War are increasingly inherited by the baby-boom generation. In 1983 only 18% of the properties had been inherited, 2.8% by gift.

Data is not available for property owned by companies because real property is normally included within the share of assets listed as technical equipment, so the available figures are too rough (Schöffel 1991: 128). In the industrial sector, however, land and property is normally owner-occupied. Company headquarters are also normally owner-occupied and smaller office units rented.

Table 1.12 Private households with real properties, 1962/3–1988[1].

	1962/3	1969	1973	1978	1983	1988
total	37.9	38.8	39.5	43.6	45.5	46.7
with regard to social position						
farmer	98.5	96.3	92.3	89.9	92.6	92.1
self-employed	60.3	65.5	67.4	70.4	72.3	73.5
civil servant	30.7	38.6	49.4	50.8	54.0	56.8
white-collar worker	28.6	34.6	37.0	46.0	48.0	48.3
blue-collar worker	32.1	36.6	40.4	46.7	49.9	48.5
not employed	30.5	30.3	29.9	32.9	35.1	39.0
with regard to size of households						
persons						
1	20.6	20.3	18.8	21.4	22.0	24.6
2	33.1	37.0	37.8	43.9	47.3	51.2
3	37.9	41.4	42.7	51.5	56.5	59.3
4	45.6	48.4	50.6	59.1	65.6	67.3
5+	59.9	62.5	63.8	70.3	77.6	76.4

Source: Wirtschaft und Statistik 4/91:279.

Note: 1/ Excluding foreign households and those with monthly income over 25,000 DM.

Financial aspects of ownership

On average, buyers of new properties have a monthly net income of
DM4,900 given that there are 1.5 wage earners per household, whereas
buyers of second-hand properties earn DM4,500 (Raschke 1991: 259). Only
25% of households with a monthly net income of below DM1,200 are property owners, while 84% of all households with a monthly net income of
between DM7,000 and DM10,000 possess real property (Euler 1991: 279).
By comparison, the average national net income is currently about DM2,300
per month per employee.

A quarter of new buildings are acquired by households with a monthly net
income of below DM3,000, and they take on monthly mortgage repayments
of below DM1,000. The average mortgage payment per month amounts to
DM1,100 for second-hand properties and DM1,300 for new buildings. Nowadays, households have to raise nearly 100 times their monthly net income
in order to purchase their properties. This is the average for all types of
households (employees, pensioners, students, etc). Households of employees
need approximately 80 times their monthly net incomes (GEWOS 1990b: 11).
The relatively low rate of capital burden per month can be explained by the
high proportion of equity capital (approximately 40%) normally contributed
from savings by the purchaser.

Private properties are currently estimated to be worth between DM3.1 and
DM3.6 billion. The *Einheitswert* (standard value), which is based on figures
from 1964, is available. According to this old Einheitswert the average

Table 1.13 Private households with real property according to the *Einheitswert*
(standard value).

DM	households 1000's	%	Standard value average per household DM	total mln DM	%
below 10,000	860	7.5	5,938	5.1	1.0
10,000–20,000	1,815	15.8	15,145	27.5	5.4
20,000–30,000	1,955	17.0	24,842	48.6	9.5
30,000–40,000	1,986	17.2	34,930	69.4	13.5
40,000–50,000	1,548	13.4	44,741	69.2	13.5
50,000–60,000	1,308	9.0	54,505	56.6	11.0
60,000–80,000	1,100	9.5	68,611	75.5	14.7
80,000–100,000	521	4.5	88,443	46.0	9.0
100,000–150,000	372	3.2	122,243	45.4	8.9
150,000 +	335	2.9	205,719	68.9	13.5
Total	11,529	100	44,430	512.2	100

Source: Wirtschaft und Statistik 4/91:280.

Note: Excludes foreign households and those with a monthly net income above DM 25,000.

wealth in properties per private household amounts to nearly DM45,000. The estimated market value comes to DM290,000 (Euler 1991: 280).

Table 1.13 shows the situation of residential property ownership in Germany in respect of the Einheitswert. The Einheitswert increases in proportion to number of persons per household and to monthly net income. Property belonging to self-employed people has the highest average value (about DM77,000), while that of unemployed persons is only DM33,000. Of households with real property, 7.5% have a standard property value of below DM10,000, while 2.9% of households have a Einheitswert of more than DM150,000. Nevertheless this small group represents a share of 13.5% of the total amount of all Einheitswerte.

Although a high rate of equity capital is usual, 54.5% of households owning property have mortgages. The average mortgage amounts to DM94,000 per household, a total of DM590 billion (Euler 1991: 281).

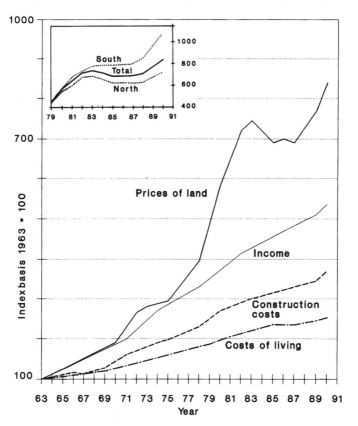

Figure 1.6 Trends in land prices for residential areas since 1963 compared with the costs of living, construction costs and the level of income.

Level and trends of prices

The land market In Germany, upward trends in land prices have always been clearly stronger than increases in the overall cost of living or in income levels (see Fig. 1.6). From 1976 to 1982 land prices increased by approximately 14.2%. Nevertheless, in the mid-1980s prices decelerated and in 1985, for the first time since 1960, they fell by 4.8%. Since then, they have risen again because of increased demand and shortage of building land.

In 1989 the average price of building land was DM126 per m². But there is much spatial variation, for example, between the cities and rural areas and between south and north Germany. In 1988, the price of building land in metropolitan areas was twice that in rural areas (DM178, compared with DM89). Top prices of up to DM345 were paid in inner-city areas, especially in München and Stuttgart as the respective Valuation Committee reports show, and nowadays prices of up to DM1,500 per m² are not exceptional (Stadt München 1990, Stadt Stuttgart 1990a).

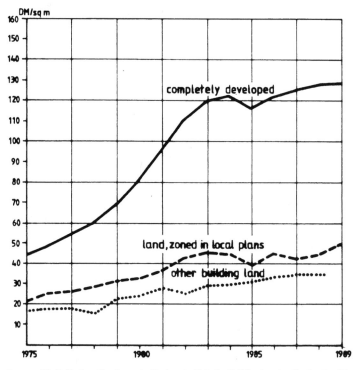

Source : Statistisches Bundesamt : Fachserie 17, Reihe 5 "Kaufwerte für Bauland"

Figure 1.7 Average values for building land.

The property market Price trends in the property market are similar to those in the land market. Since 1988 property prices have increased. Interest rates have stayed high, but demand for houses and apartments has not decreased.

The prices of single-family houses have continued to increase most rapidly. Demand for second-hand family homes has increased, making them an important sector of the housing market. This is partly because of the high construction costs of new houses and the need for immediate availability of property. Since 1990 the average cost of a single-family home has increased by about 12%, equivalent to twice that of the late 1980s. At the beginning of 1991 the price for a typical single-family house of 125 m² of living space was about DM450,000, while at the beginning of 1990 it was DM400,000 (VHW 1991: 51).

The highest house prices are found in southern German cities. A single-family house there costs on average just over DM600,000, with prices in München of DM850,000 and in Stuttgart of DM750,000. From 1989 to 1990 prices increased by approximately 19%. This contrasts with average prices of DM400,000 in Hamburg, DM290,000 in Hannover and DM250,000 in Bremen. The exception in northern Germany is Berlin, where single-family dwellings cost on average DM750,000 (RDM 1991). An increase in the cost of such property of 6% a year is typical in north and west German cities.

The price of apartments has shown a more moderate rate of change, increasing by about 8% in 1991. The average apartment is of 70 m², with between 2.5 and 3 rooms, and costs nearly DM3,100 per m². In the northern states the average price of DM2,000 per m² is about one-third lower. The highest prices are in Düsseldorf (DM4,200) followed by Wiesbaden (DM4,000), München (DM3,900), Stuttgart (DM3,700) and Berlin (DM3,600) (VHW 1991: 52). Table 1.14 shows the prices in the 1980s for single-family houses and apartments divided into new and second-hand properties.

Table 1.14 Prices for new and used houses.

| | new houses | | used houses | |
| | 1983–85 | 1987–89 | 1983–85 | 1987–89 |
	DM	DM	DM	DM
condominium	202,000	211,000	146,000	142,000
single family house	278,000	308,000	204,000	244,000
two family house	398,000	434,000	256,000	276,000

Source: Raschke, W.D. Der gläserne Bauherr - Zur LBS- Wohneigentumsstudie 1990 - In: Der Langfristige Kredit 8/1991:258.

Rents From 1980 to 1985 changes in the cost of renting a property have followed changes in the cost of living. During this period there was a continuous surplus of flats, which was followed by a high vacancy rate in 1984–5.

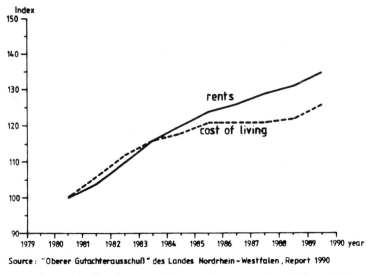

Source: "Oberer Gutachterausschuß" des Landes Nordrhein-Westfalen, Report 1990

Figure 1.8 Development of rents and cost of living 1980 to 1989.

In 1987 the average monthly rental of a flat was DM6.91 per m², excluding water and energy costs. The highest rents are paid in the centre of metropolitan areas (DM7.36 per m²) and the lowest in rural areas (DM5.87 per m²) (BMBau 1990a: 90). Higher rents are charged for brand-new flats.

Currently rents are increasing rapidly because of increased demand for housing, and up to DM12 per m² has to be paid for new flats. In prime locations in the large cities the rent for apartments can be more than DM20 per m²: up to DM27 in Berlin, DM25 in Düsseldorf and DM23 in Frankfurt and München. Normally rents in other large German cities are between DM10 and DM15 per m² for new dwellings in prime locations. Table 1.15 gives further details of rents by region, age of buildings and whether flats are financed free or subsidized publicly.

During the past 10 years the percentage of income spent on rents has increased slightly, although there are differences between the available budgets of households. In 1980 households of pensioners with small incomes had to pay out 19% of their monthly income in rent; by 1989 this had increased to 22%. By comparison, the percentage of income spent on mortgages by owner-occupiers is 5% higher (see Fig. 1.10). Households with medium incomes paid out 13% of their income in rent in 1980 and 16% in 1989. By comparison, households with a high income had to pay only 11% in 1980 and 14% in 1989 (BMBau 1990b: 27). In 1987 the average total rent per month and household was DM456 per m^2 (Winter 1991: 174).

Table 1.15 Average rent of flats with regard to age, free financed or subsidized and different spatial types of regions in 1987.

	Average	Rent per sq m per month						
		pre 1948	Date of construction					
			1949–1968		1969–1978		after 1978	
			pbc[1]	pvt[2]	pbc	pvt	pbc	pvt
Regions with large urban agglomerations								
total	7.24	6.26	6.31	7.64	7.62	8.63	7.51	9.72
core city region	7.34	6.43	6.35	8.10	7.68	9.66	7.54	10.94
highly populated surrounding countryside	7.18	5.88	6.17	7.09	7.51	8.09	7.40	9.20
less populated surrounding countryside	6.78	5.56	6.26	6.68	7.48	7.44	7.56	8.25
Regions with major urban centres								
total	6.33	5.46	6.07	6.27	7.22	6.90	7.26	7.76
core city region	6.97	6.12	6.41	6.98	7.72	8.29	7.70	9.41
less populated surrounding countryside	6.04	5.14	5.81	5.93	6.93	6.54	7.01	7.35
Rural regions (total)	5.84	4.96	5.46	5.72	6.79	6.38	6.84	7.25
Total average	6.87	5.94	6.20	7.06	7.48	7.84	7.41	8.76

Source: Wirtschaft und Statistik 3/1991:171.

Notes: 1/ pbc = public, 2/ pvt = private.

The highest rents per square metre of living space are paid for the smallest flats. In 1987, for example, flats of less than $40\,m^2$ were the most expensive at DM9.96, while flats in the social housing sector of between $40\,m^2$ and $70\,m^2$ were the cheapest at DM6.55. Since 1968, rents for small flats have become relatively more expensive because of increased demand for single-household property.

Rental costs of industrial space differ greatly and such premises are normally available only in the main growth towns. Table 1.16 shows the costs of renting on prime industrial sites in the most expensive cities. By contrast, prices in Dortmund and Köln are between DM4 and DM8 per m^2.

Data on rents for office space differ between the leading German real-estate agencies. Normally only prime office rents in the large cities based on new rental contracts are published. These can be very different from average rents. Usually the city centres have the highest values, although rents differ within inner-city locations (see Frankfurt case study, Ch. 11.2). There are also nationwide and regional differences. For example, the old industrialized "Ruhrgebiet" area has monthly office rents of between DM14 and DM22 per m^2 (Dietz & Ernst 1991), while in Frankfurt and even Düsseldorf typical rents may be DM80 and DM40 respectively (Brockhoff Zadelhoff 1991: 14).

Table 1.16 Prime rents for production space, 1989/90.

| | Monthly rents per square metre | | | |
| | Warehouse/production space | | Service space | |
	DM	ECU	DM	ECU
Stuttgart	11.50	5.5	15.00	7.2
München	10.50	5.5	14.00	6.7
Frankfurt	10.50	5.0	14.00	6.7
Düsseldorf	10.00	4.8	11.40	5.5
Hamburg	9.50	4.6	12.00	5.8

Source: Jones Lang Wootton, Consulting & Research, Market Report, West Germany, 1990.

Note: The rents quoted relate to "high-tech" accommodation, and show the top rent achievable for the best-quality buildings in prime industrial locations.

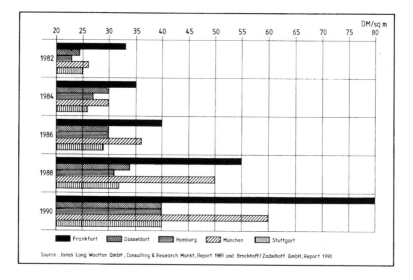

Figure 1.9 Prime office rents in DM perm^2. and month 1982–90.

In Berlin an extraordinary growth of prime commercial property rents has taken place. This has recently been accelerated by the decision of the Federal Government to locate the federal capital there. Since 1989, Berlin has had an increase second only to Madrid when comparing the index rates of all the large west European cities.

Figure 1.10 shows the strong increase in rents compared with other property prices and the cost of living in the past five years. In particular, rents for new flats have increased by 27%, while rents for refurbished flats have increased by 30%. There has also been a clear increase in rent for commercial premises.

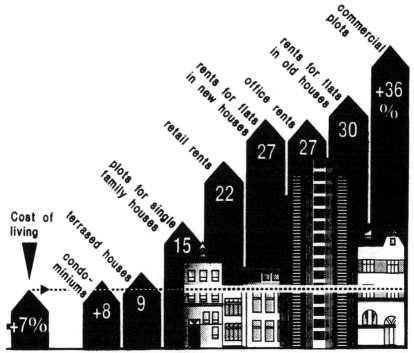

Source: Westfälische Rundschau 11/1990

Figure 1.10 Development of rents compared to other properties or forms of tenure from 1985 to 1990 (%).

Construction costs The average construction cost for a flat has increased since the beginning of the 1970s. Construction costs compared with those of the previous year decreased in the early 1980s and then increased again from 2.2% in 1988 to 3.5% in 1989. The growth rate, differentiated between conventional construction and prefabricated buildings, has shown similar trends.

Single-family houses are normally twice as expensive as multi-family houses (BMBau 1990b: 29). Expenses such as fees for architects, solicitors and surveyors amount to between 10% and 15% of total costs (see Ch. 4.2). Building costs for commercial buildings differ according to the quality of the construction required. Industrial premises are often built in lightweight materials with a concrete ceiling and simple flat roof. They usually cost between DM600 and DM700 per m^2, with complete concrete construction costing about DM100 per m^2 extra. For high-quality premises, for example,

Table 1.17 Average construction costs for new flats.

			DM		
	1970	1975	1980	1985	1989
Single-family houses	93,100	157,200	215,200	247,800	271,300
Several family houses	43,100	69,400	110,000	122,900	132,300

Source: Bundesministerium für Raumordnung, Bauwesen und Städtebau, Haus und Wohnung - Im Spiegel der Statistik, Ausgabe 1990:39.

using red brick, costs may increase to DM1,000 or more per m^2. The costs for service and office space amount to DM1,000–1,800 per m^2, depending on quality, with the current average cost of office space being about DM1,200.

Requirements for land
Land-use and land conversion   During the period 1950 to 1989 the urban area in Germany increased from 7.5% to 12.2%. Woodlands currently make up 29.8%, farmland 53.7%, and other land 4.3%. Total built-up areas have increased during this period from 3.2% to 6.2% of the overall land area, with road space increasing from 3.5% to 4.9%. The settlement structure varies between the spatial types of urbanized regions (see Table 1.18), with the cores of large urban agglomerations having the highest density. On average, about 51% of the area of a municipality is built up, although figures of up to 70% are found.

The average amount of land developed per day increased rapidly until the beginning of the 1980s. Since then it has decreased to approximately 87 ha per day. The lowest rate refers to rural regions, while the highest growth has taken place in the countryside surrounding conurbations (Table 1.19).

The decrease in the amount of land conversion can be partly explained by increasing land values, which have forced owners to buy smaller plots (see the Stuttgart case study, Ch. 7.2) and by the public policy in the 1980s of giving priority to the use or re-use of land reserves and of protecting the existing environment (*Innenentwicklung*) (see Ch. 2). The decline in the construction sector and the demand on the market, however, appear to be the main reasons. It remains to be seen whether the construction industry will produce another increase in land conversion or whether the policy of the Innenentwicklung will limit it.

32

Table 1.18 Development of land conversion for buildings and traffic in Germany, 1981–5 and 1985–9, subdivided in spatial types of regions.

Spatial type of region	1981–1985			1985–1989		
	1000ha	ha/day	$\%^1$	100ha	ha/day	$\%^2$
regions with large urban agglomerations						
core city region	16.2	11.07	4.44	12.3	8.40	3.23
highly populated surrounding countryside	28.5	19.51	6.53	24.8	16.98	5.34
less populated surrounding countryside	18.5	12.7	5.56	13.3	9.08	2.26
sub total	63.2	43.27	5.57	50.3	34.46	4.2
Regions with major urban centres						
core city region	3.5	2.38	3.73	3.5	2.42	3.65
less populated surrounding countryside	58.5	40.08	6.59	40.7	27.89	4.3
sub total	62	42.45	6.32	44.3	30.31	4.24
Rural regions	40.2	27.55	6.3	31.9	21.84	4.7
Total (FRG)	165	113.28	6.01	126.5	86.62	4.33

Source: Rach D, Erste Ergebnisse der Flachenerhebung 1989 - Trendwende im Flachenverbrauch, in Mitteilungen der Bundesforschungsanstalt fur Landeskunde und Raumordnung 5/90:3.
Note: 1/ 1985 compared to 1981 in %.
2/ 1989 compared to 1985 in %.

Table 1.19 Average daily rate of land conversion.

Period	Average land conversion per day in ha
1950–55	18
1955–60	67
1960–65	95
1966–70	138
1970–75	94
1979–80	167
1981–84	114
1985–88	87

Source: Wirtschaft und Statistik 6/1990:393

Living space per person The average living space per person increased by nearly 50% from 23.8 m² in 1968 to 35.5 m² in 1987 (BMBau 1990a: 90). There are no important differences between the conurbations, surrounding countryside or rural regions, but there has been a slow deceleration of growth in living space since the beginning of the 1980s.

The amount of living space per person varies according to the number of persons per household and also according to whether the property is rented or owner-occupied. The average space for owner-occupier households is nearly 38 m² per person, while tenants have an average of 33 m². Single households have about 82 m² if they are owner-occupiers, and 55 m² if they are tenants. Households with four persons occupy 37.7 m² (owner-occupiers) and 26.7 m² (tenants).

The average size of flats increased from 71 m² in 1968 to 81 m² in 1978 and 86 m² in 1987. Nevertheless, since the 1980s the yearly growth has decreased. The reasons for this are the decrease in the number of persons per household and slow economic growth (see Ch. 1.3) Across the regions, the average size of flats varies: in the inner-city areas it is 72 m², while in the rural areas it is between 89 and 97 m² (BMBau 1990a: 89).

Table 1.20 Living space per square metre per person, (old FRG).

	Owner occupier	Tenant
single house-hold	82	55
two- family household	49.5	35
four- family household	37.7	26.7
five and more-family house-hold	30.8	21.8

Source: Compiled from Das Haus 6/1990

A useful indicator of the general living conditions of households is the number of persons per room. In 1968, 11% of all households still had less than one room per person, while in 1987 this was true of only 2.9% of households (BMBau 1990a: 91).

There is a trend towards more individual living space because of an increase in the average income. A reduction in the growth of living space per person is to be expected, however, because of scarcity of flats, the absence of the anticipated demand for new flats in the new states of the former DDR, and the increased immigration from eastern Europe.

Space per person in the workplace It is difficult to get current data on space per person in the workplace. In the production sector it differs substantially between the types of branches. Furthermore, it has changed because of changed production systems, such as automation. This means that there are problems in calculating demand for workspace (Bauer & Bonny 1987: 43). Sometimes generalized indicators, often with regard to the number of employees per company, are used. This gives a figure of between 150 m^2 and 300 m^2 of floorspace per employee (Bauer & Bonny 1987: 59).

It is not clear whether demand per employee for floorspace will increase. On the one hand the space per employee increases because of a permanent process of automation, with the result that fewer workers require more working space per person. On the other hand, companies are continuously reducing space, for example for storekeeping, and using just-in-time production methods. It has been suggested that the criteria of "employees per company" should be abolished, to be replaced, for example, by criteria that take into account the types and numbers of machines per company.

It is easier to find suitable criteria for the space per place of work in the office sector. Normally, gross floor space in square metre per employee is used. This has continuously increased in the past decade and the trend is set to continue. It currently varies between 12 m^2 and 38 m^2. If the category includes storekeeping, the range is between 10 m^2 and 170 m^2 of floorspace (Hennings 1990: 155). At present, about 25 m^2 of floorspace per employee is used as the standard rate for the purpose of estimating the demand. If the working place is equipped with computers and telecommunication facilities, up to 20% extra floorspace is needed.

1.5 Trends in spatial development

National trends

Because of its historical development, Germany has a rather polycentric or decentralized spatial structure compared with other European nations such as France or the UK. Germany has a tiered system of towns with upper, medium and basic status. This situation has not changed since unification, although current regional problems cannot be compared with earlier ones.

Germany currently has a total population of about 79 million, 62.6 million in the western states or old Länder and 16.4 in the eastern states or new Länder (Statistisches Bundesamt 1990). Today, only Berlin, Hamburg and München have a total population of more than 1 million: Berlin 3.3 million, Hamburg 1.6 million and München 1.2 million.

The balanced spatial structure in the old Länder is supported by the municipality hierarchy. Disproportionate development occurs only in the larger cities (Deutscher Städtetag 1990). This can be demonstrated by the yearly construction of new buildings. In western Germany in 1989 similar numbers of new buildings were built in towns with more than 100,000 inhabitants (27,272 buildings) as in those with between 20,000 and 100,000 (29,410 buildings).

Table 1.21 Sizes of German municipalities (incl. East Germany).

Population	Number of municipalities
greater than 500,000	14
200,000–500,000	24
100,000–200,000	44
50,000–100,000	98
20,000–50,000	434
Zero–20,000	15,448

Source: Der Städtetag 11/1990:821ff and Staitisches Jahr-buch 1989:607.

Although there is a comparatively balanced decentralized spatial structure, considerable variations still exist.

The development of different areas in Germany is continuously monitored by a special system of regions known as the Raumordnungsregionen. Fixed criteria for evaluation, such as income levels, unemployment levels and migration levels, are used. The western states are separated into 75 Raumordnungsregionen. Since unification the monitoring system has become more detailed. The system is now called the *Laufende Raumbeobachtung* and is carried out and published yearly by the Bundesanstalt für Landeskunde und Raumordnung or BfLR (Federal Institute of Spatial Research). The Raumordnungsregionen are subdivided according to population density into the following types of regions (see Map 1):
Regions with large conurbations, subdivided into:
○ central core areas;
○ highly populated surrounding countryside and;
○ less populated surrounding countryside.
Regions with major urban centres, subdivided into:
○ central core areas;
○ less-populated surrounding countryside and;
○ rural regions.
Spatial development can be explained very well using these regional divisions, although problems remain in the determination and later evaluation of some criteria.

Until the 1980s there was concern and discussion about the contrast be-

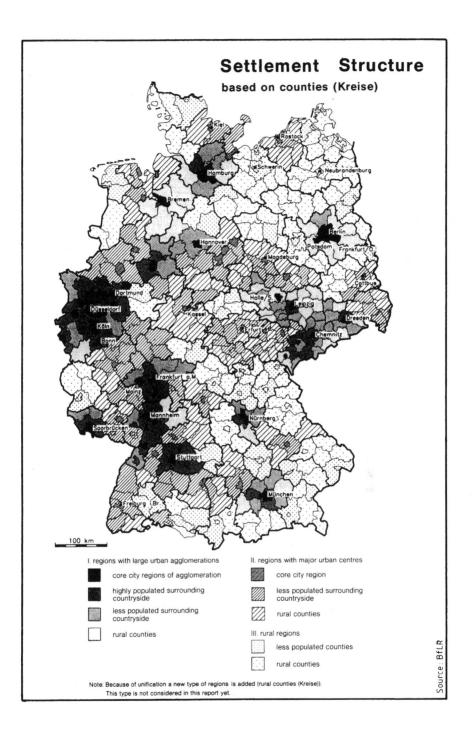

Settlement Structure

based on counties (Kreise)

100 km

I. regions with large urban agglomerations
- core city regions of agglomeration
- highly populated surrounding countryside
- less populated surrounding countryside
- rural counties

II. regions with major urban centres
- core city region
- less populated surrounding countryside
- rural counties

III. rural regions
- less populated counties
- rural counties

Note: Because of unification a new type of regions is added (rural counties (Kreise)).
This type is not considered in this report yet.

Source: BfLR

Map 1 Settlement structure.

tween peripheral and urban areas. The question of equality of living conditions and standards was then the central theme. Since then a new discussion has evolved. In addition to the traditional polarization between cores and peripheral areas, the question of regional imbalance among the different types of metropolitan areas themselves arose.

There is a national north/south divide between the northern and southern states in Germany (BMBau 1990a: 20). This imbalance has been superimposed by a new east/west divide caused by the unification and the legacy of the completely planned economy in the former German Democratic Republic (DDR). Table 1.22 explains the south/north difference using several criteria. Further criteria are land, property and rental prices, which have shown an imbalance since the beginning of the 1980s with a continuously higher level in the southern German states (see Ch. 1.4 above). Although there are visible differences between the states, the figures given describe only the average situation. In reality the whole of the south of Germany cannot be regarded as a uniformly prosperous area. Areas of traditional industry and large rural areas are often far less prosperous (Ache et al. 1989).

The main determinant for spatial development nationally is the structural change from the industrial sector to the service sector. This has been promoted especially in the southern states. New-technology industries such as those involved in communication systems and microelectronics are predominantly located in the south (Brake 1986: 171).

In future the east/west divide will determine the main national trends of spatial development. Nevertheless, the south/north divide will remain. In addition, the variations between the urban and rural areas will continue. Because of the EC's Single European Market this difference is expected to intensify and larger cities will be open to increasing competition between themselves (Ache et al. 1989: 10).

Table 1.22 Criteria for the north–south divide.

Group of states	Demographic change between 1984 and 1988 in %	Income per month and employee between 1985 and 1989 in %	Unemployment rate in 1989
north	(0.6)	17.9	9.6
middle	0.9	18.8	8.2
south	1.4	22.3	4.6

Source: Bundesministerium für Raumordnung; Bauwesen und Städtebau (ed.): Raumordnungsbericht 1990, Bonn 1990: 21 ff.

Trends in urbanization

Until recently west German cities have been mainly influenced by the process of suburbanization. The spatial structure has developed into a dispersed pattern. The process of suburbanization has, however, not occurred everywhere with the same intensity.

Between the Second World War and the end of the 1950s there was a concentration of population in the cities and central urban areas. At the beginning of the 1960s decentralization began. In the 1970s the loss of inhabitants from the inner-city areas was so high that the cities went into economic decline. It was spoken of as a process of "disurbanization" in west Germany. This process of disurbanization slowed in the 1980s and the inner urban areas were able to stabilize, with some population returning to these areas. Currently there is still a general movement out of the inner-city areas into the suburbs.

Suburbanization can be explained by different determinants that combine to form "push" and "pull" factors. These become muddled and it is difficult to separate the reasons and the causes of the trends (Arras 1979, 1983). This is especially true for land or property with high prices. Increasing pollution in the inner urban areas also pushes the population away from the cities. The extensive use of land for producing goods and the development of large satellite towns must also be considered. Meanwhile, the migration of people out of the inner areas has been followed by a similar movement by companies. Because of these trends in the 1980s, cross-links between satellite towns and settlements have grown, independent of the major centres.

With regard to the suburbanization process, the commuter belt area has continuously increased, and the number of commuters remains at a very high level. For example, out of the 7 million employees in Nordrhein–Westfalen there are about 4.7 million commuting within their municipalities, and nearly 2.2 million commuters work in a different town from where they live (Landesamt für Datenverarbeitung und Statistik Nordrhein–Westfalen 1987). Nowadays, nearly 80% of all occupational commuters between different towns use cars, while the use of public transport systems has halved in the past 20 years despite substantial improvement in the public transport network (Ache et al. 1990). Car levels have now reached about 480 per 1,000 people, a sixfold increase since 1960.

Within the suburbanization process, the level of car ownership has changed between the different spatial areas. Initially the inner urban areas noted the highest levels, while today the highly populated surrounding countryside and rural areas of conurbations show above-average levels (Landesamt für Datenverarbeitung und Statistik Nordrhein–Westfalen 1987). With regard to the supply of land and land prices, car ownership and its regional distribution is a very important factor in spatial development.

The dispersion of the city is due to several factors, such as high land costs and rent prices in the centre, new production systems and changing social values. Locations with good transport connections such as motorways or high-speed train stations will be especially preferred.

Inner urban development

After the Second World War most German cities were reconstructed according to the model of "functional city development". Many of them lost their individual and visual character. Later suburbanization processes and structural change in the economy led to inner-city districts losing much of their population. At the beginning of the 1960s, more and more low-standard housing districts in the inner cities were occupied by an increasing number of foreign workers, and social change followed. Nevertheless, there remained a net movement away from the city areas.

By the 1970s several renewal programmes were beginning to stabilize the inner-city districts (see Ch. 2). In the 1960s and 1970s, however, these pro-grammes more often followed large-scale redevelopment concepts than detailed reconstruction. This had the effect of reinforcing the trend to a mono-functional use of the inner city. The service sector began to predominate, as only office or retail companies were able to pay rising land, property and rental prices (Cihan et al. 1990). The use of inner-city land for residential purposes is no longer profitable except perhaps where expensive penthouse flats are concerned, and such areas are often unpopulated.

The 1980s halted the decline of inner-city districts and showed the first signs of regeneration. At the moment people speak of a new urbanization or even re-urbanization of the cities (Häußermann & Siebel 1987). In the mid-1980s the inner-city areas did not suffer more loss of population than the surrounding countryside, so it may be concluded that a stabilization of inner-city areas is occurring. Meanwhile, because of increasing immigration, mainly from central and eastern Europe, the conurbations are growing again and the inner cities are not lagging behind this general trend (see Chs 1.3 and 11.3).

CHAPTER 2

The policy environment

The policy environment in which the land and property market operates has changed several times during the past few decades. A brief overview is given here of the most important aspects of this since the 1950s. Five main phases can be distinguished (Zinkahn 1982: vii; Zinkahn & Söfker 1990: vii).

2.1 The period of reconstruction (1950s to early 1960s)

The situation in Germany after the Second World War is well known. Most cities had suffered serious destruction, many of them losing up to 60% of their buildings. In the early 1950s there was a housing shortage. Planning was not the first priority, as immediate practical action was demanded.

The land values had been fixed at the price levels of 1936. In the 1950s there was no federal legal base for the rebuilding of the towns, but in each of the Länder there was a planning law called the *Aufbaugesetz*. The new Federal Planning Act, the *Bundesbaugesetz* (BBauG), was long discussed and came into force on 1 January 1960. At the same time, price controls on land were repealed.

The first drafts of the BBauG in the 1950s included the instrument of a betterment levy (*Planungswertausgleich*). In 1960, however, political constraints hindered the approval of this instrument (Dieterich 1991a: 265). Another instrument that came into force together with the BBauG was a specific property tax, the *Baulandsteuer*. It was a high tax on hoarded building land. However, this tax was cancelled after four years (Ernst 1984: 371).

2.2 The period of enlargement of towns (1960s to mid-1970s)

In general, the period from the 1960s to the mid-1970s saw the planning and construction of large new residential districts in the outskirts of major cities with high population densities. These developments were often managed by large public housing associations such as *Neue Heimat*. The governments in all three tiers of the state were strongly involved in these developments, either as landowners or by providing subsidies for low-rent housing. The supply of flats was very high by the early 1970s.

In this period landowners made very high financial gains, and there was much speculation in land. Prices increased greatly. The policy changed when, at the beginning of the 1970s, the government disengaged itself from constructing flats in high-rise buildings. In 1973, before the economic recession, the number of new flats constructed per year was the largest ever recorded in Germany before the 1990s.

The major programme of construction of new flats disregarded the older housing areas in the inner cities and intensified suburbanization. At this time also, many companies expanded and looked for new locations on the outskirts of the conurbations.

The urban renewal legislation known as the *Städtebauförderungsgesetz* (StBauFG) had been discussed since the 1960s and came into force in 1971. Apart from laying down procedures for redevelopment, this legislation also covered the development of new city districts, known as *Entwicklungsmaß-nahme*, or "urban development measures". It succeeded in including an instrument of betterment levy called *Ausgleichsbeträge* for areas of renewal and areas for comprehensive development. This instrument still exists in the Baugesetzbuch (see Ch. 8.1).

2.3 The period of city regeneration (mid-1970s to early 1980s)

The period of regeneration that began in the mid-1970s is still continuing today. The economic recession in the mid-1970s and the pessimistic forecasts concerning the demographic trends resulted in vacant flats in the cities and in the new districts on the outskirts (Arras 1981: 215). The inner cities attempted to increase their attractiveness by constructing pedestrian areas in the main retail streets. Office-users moved into the inner cities because the high prices of land and property there excluded other types of companies.

In 1976 there was an amendment to the BBauG. Again an attempt was made to introduce betterment levies for planning gains. The bill, however, did not succeed. This was the last time before today that fundamental changes in land and planning law had been the subject of public discussions (Dieterich 1991b: 265). Nevertheless the amendment of 1976 extended the instruments for planning and those for implementing plans for local government and improved citizen participation in the planning process. In this period the control of rents was strengthened.

In 1977 there was an important change in tax policy. Until this time tax advantages in the housing sector had only been available to the owners of new houses. They were extended to include the buyers of existing buildings. This supported the renewal of the cities and had the effect of increasing the level of ownership.

During this period the regeneration of the rural regions (*Dorferneuerung*) was given priority. This aimed to improve living conditions and stop migration of people to urban areas. The largest structural changes in agriculture since the 1960s were the reason why Dorferneuerung was necessary and were answered by new provisions of the amended Land Consolidation Act in 1976.

In the industrial sector the policy of protection against noise and air pollution became more important. The industrial locations near to residential areas became problematic and companies had to invest in order to protect the environment. Grants were given for such environmental improvements.

The amount of public finance available decreased during this period and this meant cuts in the public sector. The amount of development of new residential and industrial land declined and subsidies were reduced.

2.4 The period of inner urban development (*Innenentwicklung*) – 1980s

Inner-city development is still continuing in the 1990s, but on a more moderate scale. The period of renewal and of scarce budgets continues. An important additional factor is the ecological movement. A programme of soil protection (*Bodenschutzprogramm*) was introduced in 1985 that aimed, for example, at the re-use of derelict land before new greenfield sites were developed. The main issues in this period were (Güttler 1987: 61):
○ renewal of towns and villages;
○ redevelopment of derelict land;
○ infilling in built-up areas;
○ solutions for mixed-used areas (industrial and residential use) and;
○ cost- and layout-efficient buildings.

The legal instruments and their application to such issues were discussed in connection with the amended planning legislation, the *Baugesetzbuch* (BauGB), which came in force into 1987. The BauGB includes both federal acts, the former BBauG and the StBauFG. Some instruments changed slightly, but in general the new law did not offer any new radical policies. Within the field of environmental policy, however, the legal regulations have become more strict since the end of the 1970s.

The financial conditions in all tiers of government were unsatisfactory until the mid-1980s. In general federal government policy consisted of withdrawal from subsidies and a reliance on free-market conditions. Thus the federal and states governments withdrew from social housing and the Länder reduced their efforts. The common financing of the renewal activities also declined.

There were few new housing developments during this period. Municipalities tried to handle the conversion of natural land resources restrictively, overlooking the fact that demand would also increase as soon as incomes increased again. In the field of industrial land more new development took place on greenfield land. The municipalities tried to redevelop derelict industrial sites, especially in the older industrialized regions. This often failed, however, because such sites had bad images and because of the problem of financing decontamination and the removal of hazardous waste. Although there are many grants available for such work, it has been more successful in the southern German conurbations where industrial land is scarce and therefore commands high prices.

On the other hand, Länder policy stressed landscape planning and environmental protection in this period. In Nordrhein–Westfalen, for example, the aim was to protect up to 10% of the land area. This issue also reduced the sites that could have been used for new developments near the cities. Nevertheless the protection of the landscape is a product of the ecological movement and a necessity for improvement in the quality of life in the growing conurbations.

There has been a concentration in the ownership of land and property by individuals and large corporations. The aim of raising the level of owner-occupation has made little progress. In the rural regions the concentration of land and property ownership is a result of the structural changes in agriculture. In the housing sector there was limited success in raising the number of owner-occupied flats by offering tax advantages. In 1987, however, there was a policy change that made owner-occupation less attractive (see Ch. 4.3).

2.5 The current period (1988–)

The current period is still characterized by the problems and policies described in the last two periods. In addition, new greenfield development has become more significant. In 1990 Nordrhein–Westfalen cancelled a ministerial decree and prevented further development of building land for new villages of up to 2,000 inhabitants. Environmental protection is one of the most important issues. The period is influenced by the increasing economic power of Germany since the mid-1980s, but also by the increasing immigration caused by political changes in central and eastern Europe. Thus the problems and policies are many and diverse in the current period.

In 1990 the federal government brought into force the *Wohnungsbauerleichterungsgesetz* (housing development law). This legislation was intended to intensify some existing instruments and reintroduce some former instruments. In addition specific regulations applicable to the new Länder in eastern Germany came into force (Ch. 3). Previously, eastern Germany had no consistent land policies.

Policy is now concentrated on the financial instruments and incentives. The tax advantages in the housing sector have been greatly extended and the government is again engaged in low-rent housing programmes. In the commercial sector the activities of the government will be concentrated on the new Länder for the next few years and so in west Germany grant allocations have declined for projects such as urban renewal.

In general the main current policy objectives are:
○ provision of more housing;
○ encouragement of owner-occupation;
○ an increase in the availability of derelict industrial land throughout the country and protection of greenfield sites, but with less land conversion;
○ protection of the environment and;
○ encouragement of public–private partnerships.

This chapter has described policies directly affecting the land and property market. But the influence of such policies is often limited, since living conditions in towns and the conditions in which the land and property markets operate are influenced even more by economic and social factors. So government policies concerning such matters are at least as important as specific policies on land and property when looking at the influence of the state on those sectors.

CHAPTER 3
The market situation in
the new eastern German states

The transformation of the former German Democratic Republic (DDR) from a centrally planned command economy towards a market system has proved to be very difficult, especially as far as land and property is concerned. Under communist ideology land was thought to be without value, so new categories of ownership and property had been developed. There was "socialist ownership", divided into "ownership of the people" (which meant ownership by the state), ownership by co-operatives and ownership by social organizations such as political parties or labour unions. Personal property was only allowed for the personal use of the owner. A plot of land could therefore only be used by a private person for recreational purposes or for construction of a house for his own use; letting it to someone else was not permitted.

All industry had been expropriated and nationalized – larger companies in the late 1940s when the territory was still under Soviet administration, the rest by the DDR authorities. In 1972 the last expropriations took place, and so all industry was the "property of the people" by that time. Only a few small traders, such as plumbers and bakers, succeeded in keeping their private businesses, usually as one-person firms.

The state also took over most of the housing stock. Most buildings with more than one dwelling were expropriated. Rents were so low that building repairs could not be paid for from this source of income. When a house fell into disrepair the state often repaired such buildings (provisionally), put a compulsory mortgage on the property and, since the owner was unable to pay this off, then took the property away, giving a small sum in compensation. Thus by 1989 most blocks of rented flats belonged to the state.

There were also no private farmers left. Farms and estates of more than 100 ha were expropriated in 1946. Smaller farmers had been coerced into joining agricultural co-operatives, but many of them or their heirs remained owners of the land they brought into the co-operative. Small farmers who had obtained a plot of land when the large estates of the nobility had been confiscated, subdivided and distributed also had to join the co-operatives. If

46

they ceased to work on the land, they often lost their title.

3.1 The special situation in the new Länder

Germany has been one country again since October 1990. The process of reunification led essentially to the same legal framework being imposed on the whole country. But whereas in most aspects of daily life only a few remnants of the communist economic system of the DDR remain, and only a few special rules have been set up for the eastern Länder by the Treaty on German Unification (Einigungsvertrag), the past is not yet over with respect to land and property. Not only are there special rules for land-use planning in eastern Germany, but also whole legislative instruments containing rules for the restitution of property to former owners (or their heirs) who had been expropriated. Furthermore, other legislation regulates special investment in the new Länder. Under these circumstances a regular market in property and real estate has not yet had the chance to develop.

The economic situation

In 1990 many Germans believed that it would not take long to bring the new Länder up to the standard of western Germany. However, the state of euphoria is past. The economic deficit is larger than expected in eastern Germany, and its gross national product is still only 37% that of the western Länder. It will take at least 10 years before equal economic conditions are established.

Planning law

The unification treaty adds a new Article (§246a) to the Baugesetzbuch (BauGB) that contains special planning regulations, mostly relaxations for the east of the requirements in force in the west. It is, for example, easier to set up a Bebauungsplan without the existence of a Flächennützungsplan. Some development can be permitted with fewer prerequisites. If a right to develop according to Art. 34 of the BauGB is repealed by a new Bebauungsplan, no compensation has to be paid. Most importantly, contracts between local authorities and developers are encouraged much more than in the western Länder. The instrument of the "*Vorhaben-und-Erschließungsplan*" is a good example: a developer who is willing and able to pay for the services necessary may submit a plan of his own to the municipality. The municipality can accept the plan, with the consequence that development in the area is permitted according to this plan, rather than subject to normal planning regulation. However, these rules of public law cannot be used as often as is desirable because of shortcomings in the field of private law.

47

Private land law

For land and property the Civil Code (BGB) of 1900 is also in force in the new Länder. However, there are some specialities, and the basic rules are not familiar to the public because communist law had also governed real estate.

In the DDR the state had often conferred the right to use a piece of land that "belonged to the people" on a private person for the construction of a one-family house for private use, without conferring ownership. Usually ownership of land and buildings on such land cannot be divided, but in these cases land and buildings may have different owners. A new Sect. 5 of Art. 231 of the legislation for the introduction of the Civil Code (Einfuhrungs-gesetz zum Burgerlichen Gesetzbuch, 1896) determines that houses construc-ted on such land shall not be part of the land but shall be special property with its own entry in the land register. And, since the land register was not kept well in the DDR, Sect. 4 of Art. 233 of the Einfuhrungsgesetz zum BGB determines that this should be so even if there had not been an entry in the land register with regard to the ownership of the house before 1989. It is not yet clear how land and buildings in these cases will be brought together again.

Rent control in eastern Germany is still very strict. Only gradually will rents become high enough to allow houses to be kept in good repair. At present the condition of the housing stock is still deteriorating.

Restitution

The Treaty on German Unity contains a clause determining that expropri-ations carried out under the laws of the Soviet military administration shall not be undone. The German constitutional court upheld this clause. So the state is owner of all land and property held by the former DDR as property of the people. Most of the industrial estates and much agricultural land is still administered by the Treuhandanstalt, a body set up to privatize most of this real estate. Many of the industrial estates have been privatized, but not all.

Expropriations carried out by the authorities of the DDR since 1949 are valid if the owner has received the legal compensation that all DDR citizens were entitled to claim. On the other hand, if no such compensation was paid, the general rule is that there is a claim for restitution if the expropriation was carried out to implement state economic policy of abolishing private enter-prise and private ownership of the housing stock of blocks for rent, or if somebody lost his property because he had fled the DDR. In these cases, former owners are entitled to get their property back. In the cities about 50% of the housing stock is claimed by old owners, 80% of whom live in the western Länder. Details can be found in legislation covering the regulation of open questions about ownership (*Vermogensgesetz*). This legislation is a rather complicated set of rules and exceptions, made to do justice to the people who lost their property during the period of communism rather than taking into account the problems of the present and the future.

Members of agricultural co-operatives can claim their land back from the co-operatives and also sell their share under special legislation governing these processes (*Landwirtschaftanpassungsgesetz*).

Investment in land and property

The German government knew that it might be difficult to invest in real estate when many parcels of land and properties had claims for restitution pending. Legislation on special investment in the former DDR (Investitionsgesetz) was therefore enacted, became part of the Treaty on German Unity and has been amended twice already; a third amendment is to be expected soon. This legislation should enable difficulties arising from the general claim to restitution to be overcome. Someone who already utilizes a piece of real estate or wants to buy property in order to safeguard or create new jobs shall be entitled to buy even if the former owner asks for restitution. In these cases the former proprietor is only entitled to compensation. "Investment shall have the right of way" is the intention of this legislation.

The regulations about restitution and investment are unfortunately not very clear and are difficult to handle. In spite of the fact that not many properties have been given back as yet and not too much has been sold, and that it is mostly public authorities that are entitled to sell real estate that was the people's property, it is very often far from clear who has the better claim. This is the case for the majority of dwellings and also for many middle-size and smaller industrial estates.

The situation of private ownership

There is a merciless fight in the new Länder for real estate between those who suffered expropriation (or rather, in many cases, their children and grandchildren) who never imagined that they might get these properties back, and the present users of land and buildings, often people who bought the properties from the DDR government in good faith and therefore believe themselves to be the rightful owners. But even former owners whose claims are not contested by other private persons are annoyed – or even full of despair – about the slow procedures pursued by an inept administration that they have to endure if they wish to pursue their claims to the end. The first findings of the courts in these cases are now being published and experts are estimating that it will take at least 20 years to settle all the claims arising out of the not very skillfully drafted legislation on the matter.

The question of ownership is clearly answered only for early cases of expropriation of industry and country estates, though not for all of them. More than two million claims to restitution (97% of those raised) are yet to be decided.

Unsolved questions of ownership are the largest hindrance for people and firms willing to invest. It is estimated that more than DM50 billion have been made available for investment in eastern Germany, but these funds cannot be used because of the problems of ownership. It is still the case that purchase of property often involves considerable risk of losing the site again or at least being involved in long litigation.

Thus the supply of land and property that can be bought without risk is low. Parts of Berlin, Brandenburg, Mecklenburg–Vorpommern, Sachsen, Sachsen–Anhalt and Thuringen still have a land economy in transition. Environmental problems and widespread industrial pollution add to the problems, as do the considerable areas that have been used by the military and are no longer needed for this purpose. All these areas await recycling and re-use for urban functions.

3.2 The market

As the supply of uncontested real estate in eastern Germany is so low, difficulties might be expected in the land and property market. Indeed, problems arising directly from the demand for land and property are enormous.

Demand

Despite continuing migration of people from eastern to western Länder, where salaries are still much higher, the demand for housing in the east is high. There is not only the enormous task of repairing and modernizing the existing stock, but also need for new housing. The DDR could never provide the volume of housing demanded by its inhabitants. Statistics that seem to prove the contrary are not really reliable, and there is still a long way to go before housing conditions in the eastern and western Länder are equalized. Many new houses have to be built, for rent as well as for owner-occupation. There is a considerable potential market, but land without claims for restitution on it has to be provided.

The demand for offices is even higher, or at least more apparent and pressing. The economy of the DDR had functioned without the many private professionals, financial institutions, insurance companies, hotels and other businesses that make up the service sector in free-market economies. The service sector is growing very fast now in eastern Germany. Office space especially is in short supply. There are still examples of banks renting a few rooms in the ground floors of hotels, and offices of local authorities found in small apartments. It takes time to build new offices, and no uncontested land is available in the cities. Therefore office rentals are high, as high as in cities in the western Länder, in spite of large differences in quality. Additionally, land for industrial investment is often hard to find, and if found, bureaucratic obstacles to development are burdensome and time-consuming.

Prices and valuation

Since the general conditions in the east are so different from those in the west, many experts are pondering whether the usual definition of market value can be used in the new Länder. It is accepted that market values can be arrived at only if certain preconditions exist: if negotiations between seller and buyer can take place, if the owner wants to sell, if there is enough time for negotiations, if prices are stable during the negotiating period, if there is enough supply, and if offers influenced by special interests are not considered. Often these preconditions do not exist. But using another definition of market value will not be helpful. There is a market, though it operates under difficult conditions. They include scarcity and high demand, so the market value will be determined mainly by these factors.

Prices for land and buildings differ widely. The federal statistical office published for the last quarter in 1992 only figures for 1989 sales of building land. In that period, 18.7 million m^2 were sold at an average price of DM10.82 per m^2, with raw building land having an average price of DM10.32 per m^2, and building land ready for construction averaging DM13.94 per m^2, prices differing between DM0.09 and DM978.26 per m^2. So the statistical average is no reliable clue to the real price. In a town near Berlin, prices could be observed for parcels of the same quality of DM52.00 and DM700.00 per m^2, without any rationale for the difference. It can be said, however, that very low prices are usually paid to very inexperienced sellers in rural areas and that the high prices are for sales in the cities with high demand. Actually the prices are very high compared with prices in the western Länder when account is taken of the fact that the infrastructure is not as developed as that in western towns and the environment may not be advantageous. Land prices in the inner cities of, for example, Leipzig or Dresden, are as high as DM18,000.00 or DM20,000.00 per m^2. And if an investor looks far enough ahead, this may even be a good investment, although such prices are the cause of the shortage of land that does not suffer from claims for restitution. It is to be expected that prices will fall in due course. Supply will grow because of allocation of more building land by planning authorities as well as decisions on claims for restitution. Then the differences or shortcomings of infrastructure and environment will have more influence on prices.

The most reliable valuation method is valuation by comparison. But valuers will seldom have the data necessary for a normal comparison when making valuations in the new Länder. To overcome the shortage of information deductive methods of valuation are highly recommended and often valuations are concluded by deductions, the deduction being made either from prices for land of other qualities, from costs that can still be expected and, last but not least, by using factors for reductions according to the size of the town, the land on which the parcel is situated and the land-use according to planning law. Valuers often try to find the value by comparing a piece of real estate in the east with a similar parcel in a comparable town in one of the western Länder. Then the environmental differences are taken into account. For example, reductions may take account of poor traffic connections; surrounding countryside; environmental damage; infrastructure of the site, town and region; expectations for the next years; and planning conditions. Using such methods may however produce a value that should theoretically apply, without consideration of the market. But at present such methods are indispensable and they can lead to a plausible result. However, when assessing values in the new Länder, it is necessary to take account not only of the price, but also of the method used to find it. Only then will it be possible to judge the plausibility of a valuation.

Information

Information about land and property can be gathered from the land register and from the cadastre. However the land register was not well kept in the DDR. Old land registers were destroyed when estates were expropriated, and unfortunately the DDR authorities often did not bother to follow their own legally prescribed procedures, since the land was worthless anyway. The land register therefore does not in all cases give information about former owners and about rights of persons other than the owner of the plot to use the land or buildings. The procedures for making entries in the land register are still complicated, and always have to be examined to see whether there are claims to restitution. This often takes a long time. But it is to be expected that a reliable land register will be compiled before long. The cadastre cannot give information about the past for all land, but often a reconstruction of the old situation is possible and the cadastre is reliable on the current situation.

Everywhere, valuation committees have been set up and have started to work rather successfully, compiling much information about sales and prices and determining guiding values for land (*Bodenrichtwerte*). It will not be long before their work will be comparable in quality to that of their colleagues in the western Länder.

Actors

The land and property market in the new Länder is still dominated by public actors. The Treuhandanstalt has the duty to privatize the whole of industry and a large proportion of the agricultural land in the former DDR. Almost equally important are the Oberfinanzdirektionen, and authorities of the federal state and of the Länder, which manage a considerable part of the former property of the people. Municipalities are responsible for the housing stock that belonged to the state. A developer interested in a relatively large area of land may find out that he has to deal with several public actors, whereas private owners may be rare. But even public authorities responsible for huge quantities of land and property often hesitate to make a decision. It often takes a long time to conclude a deal.

Private owners, who are common in rural areas of undeveloped land and in the housing sector as owners of one-family houses, have learnt quickly that land is a valuable asset. Today they tend to overestimate their property's value. If, for example, the owner of a plot at one of the beautiful lakes in Mecklenburg-Vorpommern wants to sell, he is dreaming of the prices being paid at one of the lakes in Bayern or Schleswig-Holstein, whereas a buyer coming from western Germany is on the lookout for a bargain. Also, there is no normal real-estate market in eastern Germany, as far as the actors are concerned.

A market in transition

The land and property market in the Länder of Brandenburg, east Berlin, Mecklenburg-Vorpommern, Sachsen, Sachsen-Anhalt and Thuringia still lacks transparency. It is hard for the uninitiated to penetrate the rationale and mechanisms of this market. Of course, in principle nobody is barred from it. Foreign investors especially are invited. But it is still difficult for purchasers to find the property they are seeking and they may be uncertain about the prices being asked. A good advisor is necessary; it is also quite common today to use legal advice, and the advisor should be carefully selected.

It is a market in transition. The special situation created by the fall of communism and the reunification of Germany will last for some time. The intransigence of some of the public actors on the market will not disappear quickly and private actors also need time to adjust to the new conditions. But as soon as the economic prospects of the new Länder become clear and, very importantly, if the restitution of property is making progress, the normalization of the land and property market will also make progress. It is to be expected that the conditions in the northern Land of Mecklenburg-Vorpommern will become similar to those of the western Länder in northern Germany (Niedersachsen and Schleswig-Holstein), whereas the situation in the southern Länder of Sachsen and Thuringia, as well as Sachsen-Anhalt, will in the long run follow the pattern of the other Länder in southern Germany (Hessen, Bayern and Baden-Württemberg). Brandenburg will probably be influenced to a considerable extent by the situation in Berlin, while the rest of Brandenburg will rather follow the pattern of northern Germany. But nobody knows for sure. We have to wait and see.

PART II
The urban land market

CHAPTER 4

The planning and legal framework

4.1 The legal environment

The legal framework is a fundamental influence on the urban land market. No other comparable market sector is so tightly regulated by public rules and public intervention, which apply especially to the supply of building land.

Although in theory every individual owner in Germany is guaranteed Baufreiheit (building freedom), it is not in fact possible to construct a building or to change land-use without permission.

In the first place land-use is controlled by the planning instruments of different public authorities. This means that the type of use and the supply of land are widely influenced by planning policy and plan-implementation measures such as replotting and servicing. In effect, therefore, land-use is determined by various binding rules on landowners.

Secondly, urban development is strictly controlled by the *Baugenehmigungsverfahren* (procedure for planning permission), as each new building or scheme requires official building permission. Every application is checked to see whether it accords with development plans and other public interests.

With the above measures the public authorities can control regional and urban development (their function of allocation), closely manage the supply of building land, and influence land-price levels and distribution of property and ownership.

The planning system
The development process, from zoning land to primary use, needs to be explained in the context of the various levels of administration of the German planning system, and the possibility of intervention by a higher-tier authority on a lower tier. The process of *Bauleitplanung* (urban land-use planning) at the municipal level is the key, as it directly determines land zoning. To achieve a better understanding it is necessary to distinguish between the legal framework and the legal duties of land-use planning.

The legal framework for land policy The planning system is structured according to the tiers of government. The legislative authority for housing and land-use planning belongs to the federal government, but the Länder and the Gemeinden may enact supplementary legislation, rules and orders (see Ch. 1.1).

The federal government sets out a framework of guidelines for *Raumordnung* (spatial policy) for the Länder and Gemeinden in the *Bundesraumordnungsgesetz* (federal spatial policy legislation). The principles of this legislation are based on the objective that living conditions in all parts of Germany should be, if not equal, at least of the same quality, despite the nation's limited natural resources. Both this legislation and the *Bundesraumordnungsprogramm* (federal programme for spatial policy) emphasize, for example, that there should be an adequate supply of jobs in all parts of the country. Although this does not directly influence land-use, its principles affect subsidies for infrastructure programmes and development grants for building areas, mainly in rural and undeveloped regions.

The Länder have to transfer the nationwide aims of the *Bundesraumordnung* (federal spatial policy) into concrete form by setting up further regional and urban development policies for their areas. In this way federal spatial policy becomes binding on all the Länder. The *Landesentwicklungsplänen* or *Landesentwicklungsprogrammen* (plans or programmes for the development of the Länder) specify the aims for environmental and land-use planning in, for example, areas where industrial development has to be promoted and supported or areas in which recreation and housing should be developed. Although the types of plans and rules differ between the Länder, each normally sets out its ideal spatial structure, with a hierarchy of major cities, medium-size towns and other settlements connected by development axes. Most Länder have special plans that designate areas in which the natural environment or features of scenic or tourist interest are to be protected.

Since the Landesentwicklungspläne are passed as legislation, they constitute a statement of policy embodied in law and are therefore legally binding on the municipalities. Although Raumordnung and Landesplanung do not determine the land-use pattern directly, they do set policy objectives in a legal framework.

Most of the Länder have a special *Raumordnungsverfahren* (procedure for spatial development) that is obligatory for large individual development projects such as waste-disposal facilities, large business parks, large retail centres or infrastructure projects. It aims to ensure that development is compatible with the principles of regional policy (i.e. ensure that the scheme will not be prohibited!).

Legal responsibility for land-use planning in respect of *Regionalplanung* (regional planning) lies with the Regierungsbezirke (regional administrative districts), while that for Bauleitplanung lies with the municipalities.

Figure 4.1 shows the structure of the German planning system by illustrating the hierarchical connections between different levels.

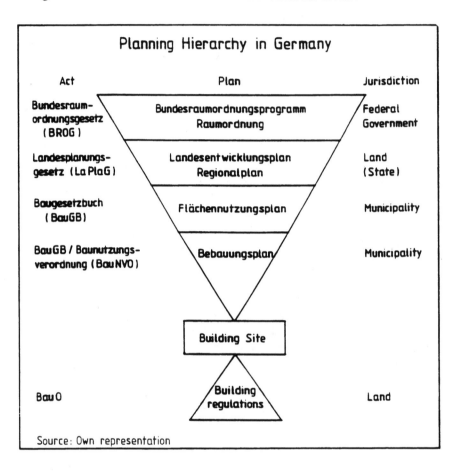

Figure 4.1 Stucture of the German planning system.

Regionalplanung adapts the aims and goals of the Landesentwicklungsplänen to specific regions within the state. The contents and organization of the Regionalplanung policies differ between the Länder, because it is their responsibility to set them. Nevertheless some generally comparable principles apply.

Generally, current and intended land-use for all settled areas, traffic infrastructure, forests, ecologically important areas and large tourist areas is laid down in maps of relatively small scale (e.g. 1:50,000). These plans are binding on every municipality. The preparation process is normally controlled by complicated procedures that involve the municipalities and thus allow them to influence the content of the Regionalpläne. Procedures vary between the Länder.

The current and intended land-uses laid down in the Regionalplänen (regional plans) arise from the general aim of a balanced spatial structure (e.g. polycentric urban structure) as well as from the expected projection of basic variables, such as the number of employees and inhabitants and their mobility.

In general, Regionalplanung sets up a broad spatial framework for urban land-use planning (Bauleitplanung) at the municipal level.

Responsibility for setting up binding and detailed land-use plans rests with the municipalities. The principles of Bauleitplanung are covered in the Baugesetzbuch (BauGB) of 1987 legislation), which is consolidating legislation, incorporating the former legislation on town and country planning and urban renewal in one statute. It is only supplemented by the *Baunutzungsverordnung* (land-use ordinance) and the new Wohnungsbauerleichterungsgesetz (legislation on facilitating the construction of housing).

Two kinds of plans fix the actual land-uses throughout a municipality:

○ the Flächennutzungsplan, the preparatory land-use plan (structure or master plan) and;
○ the Bebauungsplan, the binding land-use plan (local plan or obliging land-use plan).

A Flächennutzungsplan must be drawn up for every municipality; it shows the essentials of the intended land-use for the whole area and may not leave any "white spots". Its main purpose is to determine all types of land-use resulting from the intended urban development; it also sets the framework for detailed local plans (Bebauungspläne). As a comprehensive planning instrument, the Flächennutzungsplan has to take into account all demands for different land-use, not only demand for new building land. Flächennutzungspläne have to take account of Raumordnung, Landesplanung, Regionalplanung or other special plans (energy, environmental protection, etc.), and they have to be approved by the next highest administrative level. They are not binding for private persons or companies, but are for local authorities and for upper-level planning authorities. The latter is a consequence of the principle of local self-government of the municipalities (see Ch. 1.1). The scale of the Flächennutzungsplan is normally 1:10,000 or 1:20,000. The time horizon varies between 10 and 15 years.

The Bebauungsplan has to be derived from the Flächennutzungsplan and must not depart from it; it is legally binding on everybody. The Bebauungsplan has a more operational character and land-uses not specified within it are not permitted. It therefore offers legal certainty: by looking at the Bebauungsplan, perhaps supported by the Baunutzungsverordnung (land-use ordinance), it is possible for all concerned to tell the kind of development that will be permitted.

The Bebauungsplan determines not only the different land-uses, but also fixes several binding parameters:

○ *Art* (by type of use, e.g. housing or industry) and *Maß* (by ratio of covered area to plot area, and total floor area to plot area);
○ parts of the plots that may be covered by premises;
○ the number of floors;
○ the area for social infrastructure;
○ the area for the environment and recreation, public and private open spaces and;
○ the area for traffic infrastructure.

The Baunutzungsverordnung referred to above contains detailed and uniform rules for land-use; in other words, the Baunutzungsverordnung sets binding parameters of the Bebauungspläne. Together with the Baunutzungsverordnung, the Bebauungsplan normally determines land-use in a very detailed way, sometimes in a restrictive way (see Dortmund case study, Ch. 7.3). However, there are also examples of outline local plans.

The legality of a project cannot be disputed if the proposed development does not depart from the Bebauungsplan. On the other hand, landowners are not obliged to implement the plan. If a proposed development departs from the Bebauungsplan, exceptions can be applied, but they must be added to the local plan. It is also possible to apply for dispensation from rules of the Bebauungsplan, for example, to build more floorspace than is generally allowed. But a dispensation is usually possible only if it is to the public benefit and the interests of neighbours are not affected.

Development rights in areas without local plans Very often, large areas have been developed outside the framework of a qualified Bebauungsplan. Most of the municipalities have no intention of preparing a plan, seeing no real advantage in so doing. These areas are called *im Zusammenhang bebaute Ortsteile*, or *unbeplante Innenbereiche*: built-up zones, or zones where no binding land-use plan exists but where the surrounding area has been already built up (§34 BauGB).

In these zones a proposed development may be authorized if the project would fit into the existing built environment. This decision is based on the land-use and type of buildings in the surrounding area, and will be taken according to the rules of the Baunutzungverordnung. Furthermore, municipalities are allowed to fix the borderline between the *Innenbereich* and other areas by local law (although they do not have to), so that the delimitation will be completely clear (§34 [4] BauGB).

The potential building land supply is supplemented by these zones of im

Zusammenhang bebaute Ortsteile, but is limited by the *Außenbereich*, according to §35 BauGB (zones of open space outside the built-up areas, where in principle no new building will be allowed). It is difficult to get an application for development in the Außenbereich accepted, although it is not impossible This is because it would necessitate a change to the Flächennutzungsplan (structure plan) and sometimes even to the Regionalplan (regional plan) (see Ch. 7.3).

The procedure under §34(4) BauGB for determining the borderline between built-up areas and the open space takes less time than that for the Bebauungsplan, as there is no statutory need for public participation.

The considerable use of §34 BauGB has had a great influence on urban development. Over the past few years in some parts of Germany it has become increasingly common, for example, for medium-size business parks (mainly investment projects by private developers) to be developed without a binding Bebauungsplan, but with planning permission granted by §34 BauGB instead (see Düsseldorf case study, Ch. 11.1). This demonstrates a new tendency for the process of land conversion to become faster and more flexible. On the other hand, there is a loss of democratic involvement, which is an unfavourable effect.

Table 4.1 illustrates the importance of different instruments in granting planning permission. It shows the great influence of the Bebauungsplan on building-land supply. It can be concluded that the supply of building land in our market system is a product of planning.

Table 4.1 Importance of planning permissions in 1988.

	Total number of buildings	Local plan §30 BauGB	§34 BauGB	Agricultural land §35 BauGB
		%	%	%
housing	91,000	79	18	3
office	1,730	74	22	4
industry	12,150	66	25	9
others	7,000	27	26	47

Source: Statistisches Jahrbuch BRD 1990.

Development control by the Baugenehmigungsverfahren
(procedure for planning permission)

While the different development plans prepare for and plan land-use, the *Baugenehmigung* (building permission) can be regarded as the instrument that effectively controls development.

Each landowner intending to develop land has to make an application to the *Baugenehmigungsbehörde* (building-permission committee). This is a department of the *Kreisverwaltung* (county administration) or municipality in the case of Kreisfreie Städte. Individual landowners normally have to make an application for the following forms of development:

○ new housing construction;
○ new construction for commercial uses;
○ physical change to parts of buildings; and
○ construction of garages, etc.

The procedure for obtaining building permission is laid down in the *Bauordnung* (building code) of a state, which contains detailed rules about the technical standards of buildings, such as fire precautions and environmental protection.

The official procedure can be divided into two stages. Often the applicants first make a *Bauvoranfrage* (preliminary application) to check in principle whether it is possible to build. The Baugenehmigungsbehörde checks whether the project should be judged according to a Bebauungsplan under §30 BauGB, under §34 BauGB (built-up area without a local plan) or under §35 BauGB (open space). The Bauvoranfrage is advantageous because it saves the investor the money and time that might be lost should a full application be turned down. Use of the Bauvoranfrage is therefore recommended before buying a plot.

After the applicant has obtained a positive result from the preliminary application, an official *Bauantrag* (building application), with all details of construction, can be made.

The Baugenehmigungsbehörde checks all details, especially type and intensity of land-use, and the technical safety standards of construction. The applicant is only allowed to begin construction once written permission has been obtained. Building permission takes between two and six months to process. The building codes of the Länder normally fix the duration to two or three months, but this may be extended.

Environmental protection law and other relevant acts

The German planning system contains other pieces of legislation relevant to the process of land development. Environmental protection legislation, in particular, is becoming increasingly important.

Every Bebauungsplan involves an environmental assessment according to

EC Directive EEC/85/337, which is incorporated into German law in the national *Umweltverträglichkeitsgesetz* (UVPG, 12.2.1990). However, environmental assessment is not a separate procedure, as the *Abwägung* (weighing) process of Bauleitplanung already assesses all environmental aspects. Nevertheless, several municipalities have laid down their own rules for linking the procedures of Bauleitplanung and environmental assessment.

The *Bundesnaturschutzgesetz* (federal legislation for nature protection) (BNatschG 1987) sets up a general framework for limitation of land-uses. The main aims are to protect areas from unregulated development, to promote the wellbeing of nature and to oblige the developer to compensate for the adverse environmental impact of a project. It is up to the Länder to make these federal rules concrete by means of their own *Landschaftsgesetze* (legislation for landscape development). These enable the Länder to designate special areas where, for example, development is either not allowed or is strongly resisted (e.g. *Naturschutzgebiete* and *Landschaftsschutzgebiete*). The instruments of the Landschaftsgesetze are normally applied in the Außenbereich (greenfield areas according to §35 BauGB), where the BauGB is not often applied.

Furthermore, the Landschaftsgesetze reinforces the compensation measures for land-uses causing environmental damage. These rules vary between the Länder. Compensation usually takes the form of environmental improvement to the damaged site, but if this is not possible it can be done by relocating the affected activity or financial compensation.

Water protection is regulated through the *Wasserhaushaltsgesetz* (WHG 1990 – legislation for federal water protection) which limits water use and provides a framework for the *Landeswassergesetze* (legislation on water protection of the Länder). This legislation enables special water-protection areas to be created, in which incompatible land-use is forbidden or limited.

The *Bundesimmissionsschutzgesetz* (BImSchG 1990) regulates the damaging effects of gas, smoke, noise, smell, etc. on neighbouring properties. It states a company's duties in respect of technical production facilities, as well as what is prohibited, and is important in industrial land-use.

The *Landesimmissionsschutzgesetze* lays down regulations with specific consideration of Bauleitplanung (urban development planning). The Länder sometimes determine special *Abstandserlasse* (ordinance for distance) to protect housing areas from industrial use (for example, Runderlaß des Ministers für Umwelt, Raumordnung und Landwirtschaft des Landes Nordrhein–Westfalen 1990). There are several classes of protection, which determine the exact distance a particular type of industry should be from housing.

The planning process for local plans

The procedure of drawing up local plans is laid down in the Baugesetzbuch. In all key stages, the same rules apply to the Flächennutzungspläne and the Bebauungspläne. The process can be divided into five stages:

1. The *Aufstellungsbeschluß* (official resolution for drawing up local land-use plans) can be passed by a city council after preliminary examinations and negotiations with the parties concerned (for example, building contractors) and after the aims of the plan have been determined. The resolution must be made public. The municipality is then allowed to use the instruments of *Veränderungssperre* (order banning any physical change within an area) and *Zurückstellung von Baugesuchen* (turning down building applications) (see below).

2. The first stage of participation of inhabitants and institutions with public interests should be as early as possible. It consists of informing them of the general aims and purposes of the plan. The municipality is urged to present alternative concepts and should explain the possible impact of the schemes.

   All institutions and authorities representing the public interest have to be involved in the development. Institutions affected are, for example, the public boards for water or environmental protection, the army, the post office administration, chambers of commerce, churches, etc. All neighbouring municipalities also have to be involved. The institutions and neighbouring cities must comment on the scheme, then give their own wishes and reservations within a time limit set by the municipality.

3. The second stage of participation follow discussions of any reservations about the plans by the political and administrative committees of the municipality. After this the plans have to be made publicly available for one month. This is usually done by announcements in local newspapers and in official information papers. The actual plans and their explanatory reports are usually displayed in the town hall. All interested people are allowed to give written or verbal comment.

4. The stage of Abwägung (weighing) and approval obliges the municipality in weighing all public and private interests among each other and among themselves (§1[6] BauGB). All comments, ideas and reservations have to be taken into account in this stage of the process. Afterwards the city council may officially pass the local plans.

5. After being passed by the city council all Flächennutzungspläne have to be approved by the next highest administrative level. Bebauungspläne have only to be notified to that authority if they are derived from an existing Flächennutzungsplan. In general, the higher administrative level has to check that the plans contain no formal mistakes or breaches of law. The authority should give its decision within three months, after which

the municipality has to announce publicly the approval of the higher administration. At this point the local plans have obtained legal validity. Any changes to existing land-use plans or even their annulments follow the same procedure.

The duration of the process for drawing up local plans varies depending on the type of plan. In 1987 it was found that Flächenutzungspläne needed on average five to six years, although it can take longer (up to 13 years) especially in larger cities. By contrast, new Bebauungspläne need approximately three years to achieve legal validity (Schmidt-Eichstaedt 1987). It can be assumed that the process takes even longer today.

Private law relating to land, land transactions and first urban use
The rules of private law are fixed in the Civil Code (Bürgerliches Gesetzbuch [BGB]) from 1900. The BGB defines and regulates special types of ownership and specific forms of land-use (see Ch. 1.1).

For example, *Dienstbarkeiten* (servitudes) can be used to promote the development of land or to encourage a better land-use. In many cases, these servitudes include the right to use parts of neighbouring plots, mainly for *Wegerecht* (right of access: the right to use a neighbouring plot for access to one's own). This must be recorded in the *Grundbuch* (land register, see below). Other servitudes are, for example, the *Dauerwohnrecht* (right to live in a flat for ever) and the *beschränkt persönliche Dienstbarkeit* (special persons or institutions are allowed to use a plot or part of it in a specific way, for example for technical installations).

The servitudes based on private law have to be separated from the *Baulast* (building burden of a plot), which comes under public law. A Baulast is recorded in the Bauordnungen (building orders) of the state and typically enables an exception to the established *Bauwich* (minimum space between neighbouring buildings). It is important in inner-urban areas where a minimum Bauwich often cannot be maintained. Sometimes the necessity for a Baulast will be exploited for profit by landowners (see Frankfurt case study, Ch. 11.2).

However, in practice the most important rules in private law are those concerning burdening land and property by mortgages (see Ch. 4.2).

In Germany all transactions of land (sales, purchases) between different parties must be concluded by a written contract (Ch. 1.1). The transfer of ownership always has to be examined and signed by notaries and will be legally valid only when registered in the Grundbuch (land register).

Municipalities can influence the land-development process by using contractual real-estate agreements. A municipality can make a legally binding agreement with a buyer, for example, to require building to be undertaken within a specified period or for the right to repurchase the land. An appropriate example of the successful design and use of real-estate agreements for carrying out public land policies is the development of the Technologiepark Dortmund (Ch. 7.3).

Generally, the municipality should always become an intermediate owner of land if it wishes to implement its land policies. Developers and building constructors usually protect themselves with special opt-out clauses, in case they eventually do not get building permission.

Instruments for implementation of plans

The process of land development is facilitated by special instruments based either on private law as described above, or on public law, as described below.

Replotting of land (Umlegung) The procedure of *Umlegung*, laid down in detail in the Baugesetzbuch (§45–79), is the most important instrument in replotting land boundaries (Dieterich 1990a). It facilitates the process of the Bebauungsplan and must correspond with it. Umlegung normally has two aims, illustrated by the Stuttgart and Hildesheim case studies (see Ch. 7): to provide owners with usable or developable building plots; and to enable the municipality to take ownership of areas necessary for public development, such as streets or other public spaces. The municipality is allowed to keep the betterment levies.

Umlegung enables existing, unfavourable boundaries to be changed, if, for example, plots are too small or too narrow for building to be possible. Unfavourable boundaries often result from inheritance laws. The main aim of Umlegung is to create suitable plot structures, but it can also be seen as a method of land assembly.

Umlegungen are normally applied to greenfield developments, but can also be used in the renewal process of run-down inner-city areas. The process of Umlegung is a complicated procedure that can be subdivided into the following main stages:

1. The municipality makes the formal decision to start the procedure by determining the area of the Umlegung. At the same time a *Verfügungs- und Veränderungssperre* (special order banning any physical change) (§51 BauGB) will be passed.
2. All rights and claims of all plots belonging to the area of the Umlegung (*Umlegungsmasse*) are established and added together.
3. Land designated for streets, other public space or similar amenities in the local plan is appropriated from the Umlegungsmasse.
4. The remaining *Verteilungsmasse* (private properties) will be returned to all owners involved using a special *Verteilungsmaßstab* (standard of redistribution). There are different possible standards of redistribution:

according to plot either values or sizes. The use of the size standard is only suitable if the values of all former plots are fairly similar. The principle of allocation has to take into account the former ratio of ownership, so that if for example a landowner possessed in total 20% of the overall value of all former plots he would have to receive back 20% of the value of the reallocated plots.

5. The allocation of new plots to landowners is conducted on the basis that each gets one or more developed plots according to his entitlement, with monetary compensation if necessary.

6. When using the value-based Verteilungsmaßstab, the landowner has to pay the difference between the value of his former plot (undeveloped) and the value of his serviced new plot after the procedure of the Umlegung (betterment levies). The betterment levies are retained by the municipality. When using the Verteilungsmaßstab according to the sizes of the plots, the municipality is allowed to retain land equal to the increase in value caused by the Umlegung itself; however, according to the BauGB this may not be more than 30% in greenfield areas and 10% in inner-city locations. The former appropriation (stage 3) of land also has to be taken into account (see Stuttgart case study, Ch. 7.2).

7. The results of Umlegung are fixed in the *Umlegungsplan* and must be published. If a landowner does not agree, he can take the matter to court.

The *Grenzregelung* (§80–84 BauGB) is another instrument for replotting land. It is used to straighten out boundaries between unfavourably shaped neighbouring plots where no suitable land development is otherwise possible. The municipality handles the Grenzregelung for local plan areas and inner-city locations (§34 BauGB). The procedure is in principle similar to the Umlegung, so it is called its "little sister".

Instruments to protect planning The municipalities can use various instruments to safeguard their planning policies and ensure that the land-development process corresponds with their aims.

The most important is the Veränderungssperre (§14–18 BauGB) (order banning any physical change), which can be passed as soon as a municipality has passed an Aufstellungsbeschluß (resolution for drawing up) for a local plan. As a result, landowners are not allowed to build, change their use of buildings or demolish them for two years. Thus the freedom of action of the as yet unrealized Bebauungsplan is protected. The order can be extended for one or two years (but only in special circumstances and with the agreement of the higher administration level). After four years it is only possible to pass a new Veränderungssperre if the planning procedure of the Bebauungsplan has not been concluded and there are exceptional reasons. If this occurs, the municipalities must compensate the landowners.

The Veränderungssperre is a good illustration of the limitation of land-ownership in Germany. It can be regarded, so to speak, as the dutiful aspect of landownership (*Sozialpflichtigkeit des Eigentums*), because undesired development may be hindered without any compensation.

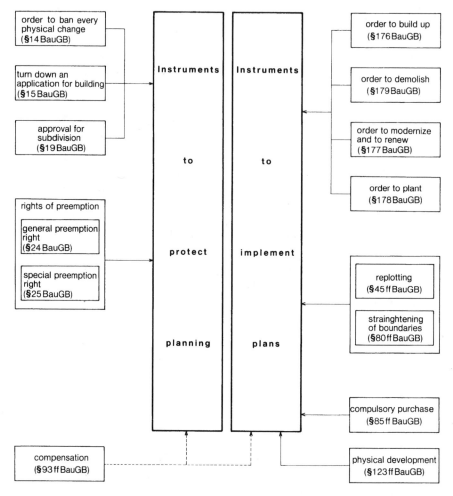

compiled to: Braam,W: Stadtplanung, Düsseldorf 1987 p. 103

Figure 4.2 Instruments for implementing and protecting plans.

A municipality is allowed to reject an application for building permission (Zurückstellung von Baugesuchen) for up to a year if the Veränderungssperre has not yet been passed by the city council (§15 BauGB). The municipality normally applies this when it recognizes the need for a local plan, as it is then required to pass an official resolution for a Veränderungssperre within the year.

Official approval is required for all subdivisions of plots lying within the area of a Bebauungsplan, a Veränderungssperre, an "inner area location" or a renewal district (§19–23 BauGB). Permission to subdivide plots is granted by the municipality, enabling them to stop subdivisions that do not promote the intended development.

If the *Teilungsgenehmigung* (official approval for subdivision) is given, the landowner is guaranteed the right to construct a new building. In principle, he has an entitlement to planning permission for up to three years. The Teilungsgenehmigung will be valid if the application has not been rejected by the municipality after a maximum of six months. After permission is granted, the land register must be modified.

The conversion of rented flats into condominiums is affected by German planning law if the flats are located in tourist areas. The conversion can be hindered if the municipality is intent on stopping too great an influx of foreigners (see also Ch. 8.1 and Ch. 11.3).

Right of pre-emption (first refusal) The principles of pre-emption rights are governed by civil law and allow each landowner to use the instrument of pre-emption in private land dealings. According to the Baugesetzbuch (public law) the municipality is able to use pre-emption rights, although only in special cases and not throughout its whole area (§24–28 BauGB and recently §3 BauGB-MaßnG). There are, in principle, two kinds of pre-emption rights:
○ the *Allgemeine Vorkaufsrecht* (general pre-emption right) under §24 BauGB and;
○ the *Besondere Vorkaufsrecht* (special pre-emption right) under §25 BauGB.

The municipality is allowed to use the Allgemeine Vorkaufsrecht only for plots designated for public use (e.g. streets, schools, environmental space) and within an area in which a Bebauungsplan is operative. It can also be used in an Umlegung-area or in a stated renewal zone. The Besondere Vorkaufsrecht gives the municipality the power to purchase plots where no Bebauungsplan is in existence, and is only valid in specific areas which the municipality can determine by local law, within areas otherwise available for normal development in the future.

The new Wohnungsbauerleichterungsgesetz (legislation facilitating housing) in force until 1995 extends the right of pre-emption, especially for greenfield locations established as new housing areas by the Flächennutzungsplan. Speculation and hoarding of building land can be prevented by the municipality becoming an intermediate owner of land. The municipality is allowed to purchase land at its *Verkehrswert* (market value), and is thus fully able to meet the demand for building land with an acceptable price.

The use of pre-emption must always be in the general public interest (§24

(3) BauGB). The municipality has to resell land as soon as the intended aim can be realized, or if the reasons for buying no longer exist (§89 BauGB), and must pay the price agreed upon by the landowner and buyer, although exceptions are possible under the Wohnungsbauerleichterungsgesetz, referred to above. By using pre-emption rights the municipality is able to obtain land in the early stages of a development and the necessity for expropriation can be avoided.

Although there are various opportunities for the use of pre-emption rights, they are not commonly used. The current legal construction makes them unsuitable for long-term land banking policies.

Compulsory purchase   The right of expropriation is laid down in the fifth part of the Baugesetzbuch (§85–122 BauGB), but is only permitted if the intended land-use cannot be achieved by any other available means, and if public welfare requires it. Serious and earnest attempts to buy have to be made first.

The public authority is allowed to expropriate land if the plots fall within a Bebauungsplan area, or if they are unused and vacant serviced sites (*Baulücken*).

Theoretically, expropriation could also be used to assemble building land for social housing, should there be a shortage. However, the rule that compulsory purchase must always be the last resort prevents its practical application. Until 1960 there was special land-assembly legislation allowing the expropriation of land for housing construction (Dieterich 1991a: 259).

However, landowners retain their right to compensation, based on the Verkehrswert (market value) of the property. Normally the Gutachterausschuß für Grundstückswerte (valuation committee) is commissioned for the valuation because it is an independent institution. Landowners may demand compensation in replacement land if they depend on land for their livelihood or profession, for example.

In practice, compulsory purchase plays only a small rôle in the assembly of building land. It has more importance in special development projects such as motorways, railways, etc. However, these are covered in special expropriation legislation set out by the Länder.

Local infrastructure development and private landowners' contributions
According to §123–135 BauGB, physical development of building land is the responsibility of the municipality.

The rules governing provision of infrastructure are vital because landowners can only start an individual development once this has been arranged. However, landowners do not have the right to force the municipality to initiate such infrastructure development. Occasionally, the municipality makes use of the possibility of employing a private building contractor to undertake infrastructure development using a special *Erschließungsvertrag* (contract about land development – §124[1] BauGB). Physical development must always follow the Bebauungsplan and an area cannot be developed without one, with possible exceptions in inner-city locations under §34 BauGB, or if the municipality has special permission from the higher administrative authority (Kreise).

Separate laws cover other areas of development, such as energy, water and motorways. For example, the *Kommunalabgabengesetz* (legislation for different fees for municipalities), which allows the municipality to distribute no more than 50% of the costs to landowners.

The municipality has the right and duty to impose a fee on landowners to cover development costs. Landowners pay a maximum of 90%, while the municipality bears a minimum of 10%. The municipality is allowed to vary the percentage level of fees and the scale of distribution for the landowners in a special *Erschließungsbeitragssatzung* (local ordinance of development). Fees follow either the real costs of development or standard prices, and are limited to public places: streets, green space, children's playgrounds and noise barriers (see Stuttgart case study, Ch. 7.2) for example. There are three methods of estimating the individual fees for each landowner:

○ according to the kind and intensity of land-use;
○ according to the size of the plot; and
○ according to breadth of the plot along the street.

The municipality decides which one is to be used. Landowners normally pay after the completion of development, but for some years municipalities have been allowed to demand fees in advance once physical development has begun (§133[3] BauGB). On the one hand, this creates pressure for selling or quick use by the landowners; on the other hand, prefinancing of development costs helps the municipality.

Order to construct (Baugebot) In areas covered by a local plan or locations under §34 BauGB, the municipalities can apply a special order urging landowners to build up their plots within a certain time. For this, two major requirements should be met:

○ justification of the importance of the urban development must be present (§175[2] BauGB) and;
○ there must be economic concessions to landowners (§175[3] BauGB) and these can demand that the municipality adopts their plot if they cannot afford the development themselves.

Municipalities face difficulties in defining important "urban reasons" and in trying to achieve equality for all landowners. These difficulties have been lessened by the introduction of the new Wohnungsbauerleichterungsgesetz (legislation on facilitating the construction of housing). As a result the definition of important "urban reasons" has been equated in law with "urgent demand" for housing. Furthermore, municipalities are allowed to combine the procedure of compulsory purchase directly with the procedure of the Baugebot. The Baugebot may be more useful due to this new legislation, although it cannot influence large-scale development projects.

Until recently, the Baugebot was rarely used, and was without great practical effect. It was only applied in a few municipalities, and only a few plots were built up using it. In many cases landowners appealed to the courts, and decisions about new buildings were thus delayed for several years.

Betterment levy In principle there are no betterment levies on increasing land values. Attempts to introduce specific rules have failed. However, some fees for the betterment of land exist:

○ The *Erschließungsbeitrag* (local fee for infrastructure development), mentioned above, enables up to 90% of development costs to be charged to landowners. Some experts regard this as a kind of betterment levy, as private landowners are financing public activities. It could also be termed an "exaction".

○ If municipalities want to profit from increasing land values within a development area, they can make use of Umlegung (replotting), as the difference between the land values before and after replotting can then be transferred to the municipality. However, part of the development costs (for acquiring land for streets, etc.) is included in these betterment levies.

○ Fees for Grenzregelung (straightening of boundaries) are of less importance, because it is not very common and often they do not cover the cost of the procedure.

○ In a *Sanierungsgebiet* (urban renewal area) (§136 ff. BauGB) landowners normally pay the difference between the land values before and after renewal. These levies must cover only the cost of physical improvement to the area. If the betterment levies exceed the real costs, the municipalities have to refund the "profit" to the landowners. However, the betterment levies are usually less than real costs and the public authorities make a financial loss. A municipality may abstain from collecting a betterment levy if the amount is too low to cover the costs of administration and estimating the levy. In estimating the betterment, only the land value, not that of the buildings, is considered. In cases where *Sanierung* (renewal) is combined with Umlegung (replotting), the betterment levy is calculated as if for replotting. Double betterment levies are not possible.

The greatest possibility of gain from increasing land values during the development process is offered by the *Städtebauliche Entwicklungsmaßnahme* (urban development measure). This instrument was less used in the past and was abolished in 1987. It was reintroduced in 1990 with the new Wohnungs-bauerleichterungsgesetz (legislation on facilitating housing) (§6–7 BauGB-MaßnG).

Städtebauliche Entwicklungsmaßnahme can be used to develop large areas for housing or industry. The municipality is obliged to buy all land in the area for a price that does not include hope value, in expectation of further development. Betterment levies are generally made possible by selling the developed building land at market value, but the municipalities must sell all plots. In this case betterment levies result from the municipality's activity as an intermediate acquirer of land.

Information systems

An important information system and a part of the legal framework is the Grundbuch (land register), referred to above. The land register is a reliable record of all real property and its owners. There are no doubts about the title to a property, because all land is registered. The Grundbuch also contains all restrictions on a site: servitudes (e.g. Wegerechte or right of access) and all mortgages attached to a plot. In this way interested buyers are always sure of the charges burdening a property. Normally, the banks base their decisions on credit on this part of the Grundbuch.

Access to the register is limited: only owners, buyers, banks or persons with a legitimate interest are authorized. All rules are laid down in the Grundbuchordnung (land registration legislation) (GBO, 05.08.1935).

The cadastre is another important land-information system, set up by the Länder. There are usually three parts:

○ Katasterbuchwerk (cadastre books);
○ Katasterzahlenwerk (cadastre figures) and;
○ Katasterkartenwerk (cadastre maps).

The Katasterbuchwerk is the most important component, recording all plots differentiated between owners, sizes, kind of use and the Ertragsmeß-zahl (basic figure for agricultural land taxation). The Katasterbuchwerk is not public, and access is only granted to people with specific interests. The Katasterkartenwerk contains all relevant Flurkarten, maps on which the actual boundaries of each plot are registered.

There are also several other public information systems, recording, for example, available building land reserves, normally on the municipal level. These so-called Baulandinformationssysteme are separated into different property sectors, greenfield developments and inner-city locations. They are not a legal requirement but are often used to work out new development plans, for example the Flächennutzungsplan.

Local valuation committees collect information about land and property values and provide information on the data necessary for valuation purposes. This is usually given in the form of land-price indices, rates of return for certain land-uses on the local market, coefficients of conversion for prices between different land-uses, etc. The material and figures published by the valuation committees are fairly reliable because all freehold, long-leasehold (Erbbaurecht) and inheritance contracts must be sent to the committee. This is a duty of notaries. It is advisable to use this source of information at least for any preliminary overall survey on land and property prices.

Each year the committees draw up and publish special maps showing actual land values. This should secure the openness of the market and ensure that buyers are better informed. The Länder determine the rules of the *Gutachterausschüsse*; they are usually affiliated to the Kreise (counties) or larger cities.

Nothing definite is known about private information systems, but several real-estate agents record different information. Information is sometimes made public, usually as market reports about major cities. Nationwide prices for land and rental levels are published several times a year by the *Ring Deutscher Makler*, a large German real-estate association (see Ch. 5.2). These figures are often used as an alternative to the public data supplied by the Gutacherausschüsse für Grundstückswerte.

Another important source of information is the *Volkszählung* (national census) and the *Mikrozensus*.

The Volkszählung contains a total count of all inhabitants, households, professions and buildings, together with their living conditions, flat descriptions and working places, and should take place every 10 years, under special law. The last one, taken in 1987, showed a substantial shortage of flats and a slowly increasing population. The data is calculated by the Bundesamt für Statistik (federal board for statistics) and relevant figures are published. Detailed figures are available for, for example, scientific work, although data protection has to be taken into account.

The Mikrozensus is the little sister of the Volkszählung. A sample of 1% of all Germans is selected and interviewed over a period of four years on the most important questions. The microcensus is often used, as the figures are available sooner, it is cheaper to carry out and public resistance is lower. The figures are also calculated and published by the Bundesamt für Statistik.

More useful data about spatial and urban development in respect of the 75 *Bundesraumordnungsregionen* is available from the Bundesforschungsanstalt für Landeskunde und Raumordnung (federal research institution) in Bonn–Bad Godesberg (see Ch. 1.5).

4.2 The financial environment

The size of the market

In 1990 investment in the construction sector totalled DM324 billion (Deutsches Institut für Wirtschaftsforschung 1991: 330), which is equivalent to 13% of GNP (the figure does not include the cost of land or plots). The building sector has grown since 1989, following a moderate decline in the mid-1980s. This recent growth was faster than the growth of GNP and in 1990 was 11% higher (in market prices) than the previous year (Strohm 1991: 20). This data, from the Statistisches Bundesamt, differs little from that collected by the Deutsches Institut für Wirtschaftsforschung, Berlin.

The building sector is divided into three categories, which have had the following shares of investment since 1989:

Housing construction	46%
Commercial and industrial construction	30%
Construction ordered by public authorities	24%

While investment in housing and public construction fluctuated during the 1980s, investment in commercial and industrial construction continued to rise. In 1975, this category's share of investment was 24%. In the 1980s the building sector was dominated by refurbishment, but in 1989 new construction regained a 50% share of investment.

Turnover in the land and property market is another indication of market size. In 1989 the value of purchase transactions in the housing sector was approximately DM160 billion, of which DM20 billion resulted from inheritance (GEWOS 1990a). Another source estimates that in 1990 transactions involving change of ownership of used properties were worth DM180 billion (Preisinger 1991: 255). Further increases are to be expected in coming years.

Sources of financing

Building construction is financed with varying shares of equity and loan capital. In Germany there is no legislation specifying the proportion of equity capital that should be involved in the financing of a building project.

The most important type of loan capital used in the construction sector as a whole is the *Hypothekarkredit*, which is a mortgage attached to property as security. The mortgage must be recorded in the Grundbuch (land register) so that potential buyers can see what charges burden a plot. There are two kinds of security for a Hypothekarkredit: the *Hypothek* (mortgage) and the *Grundschuld* (land charge, encumbrance). Although in principle there is no difference, the Grundschuld is more commonly used today since it is far easier to handle. In 1989, about 40% of all existing credits were in the Hypothekarkredit category; these long-term credits totalled DM845 billion (*Statistisches Jahrbuch* 1990: 323). In 1989, the market shares of the most important credit institutions dealing in mortgage-credits were:

Banks	80%
Mortgage banks (refinanced by mortgage bonds)	43%
Savings banks (*Sparkassen, Banken*)	21%
Building societies (*Bausparkassen*)	11%
Insurance companies	9%
Life insurance companies	8%

In the past five years, 75% of bank-held credits were in housing projects, as were 95% of insurance company credits and almost 100% of building society credits (Statistisches Jahrbuch 1990: 323). The sum of new mortgage credits in the housing sector for 1989 was approximately DM60 billion (own calculations based on *Grunddaten zur Wohnungs- und Bauwirtschaft* – Bundesbaublatt, 1989/90 issues). The other components of financing for land and property projects are explained below.

The conditions of financing

The most important factor influencing loan capital is the interest rate. Naturally this is strongly affected by the general macroeconomic climate and is determined by market forces. The conditions of the financial market are influenced by the policy of the Deutsche Bundesbank, which has the legal duty to stabilize the currency. One means by which this is done is to fix the two interest rates, the *Diskontsatz* (discount rate) and the *Lombardsatz* (Lombard rate). These are the rates at which commercial banks can borrow money from the Deutsche Bundesbank and so represent the base on which these banks in turn set their lending rates. At the end of 1987, the Diskontsatz was set at 2.5% and has been raised to 6.5% since the beginning of 1991. The Lombardsatz, which is now set at 9%, is normally set two percentage points above the Diskontsatz.

The "effective yearly interest rate" is important to the customer, since it takes account of adjustments if a loan is borrowed at discount or premium rate (e.g. 97% of loan borrowed, but paid back in full over the loan period), as well as loan arrangements and management fees, etc. Credit institutions must by law give details of this rate (Preisangabeverordnung 1985), so customers can readily compare loan offers from different banks. In the housing market, the Hypothekarkredit interest rate fluctuates at a rate representing the whole property market, as shown in Figure 4.3.

Customers may choose either fixed-interest rate loans (usually fixed for five years, but sometimes for 10 or even 15 years, although this raises the effective rate), or a variable market-orientated rate that presents the risk of sudden increases but the potential benefit of lower rates.

There are many publications that stress the serious implications of rising interest rates on the financing of a building project (e.g. Creutz 1987). The

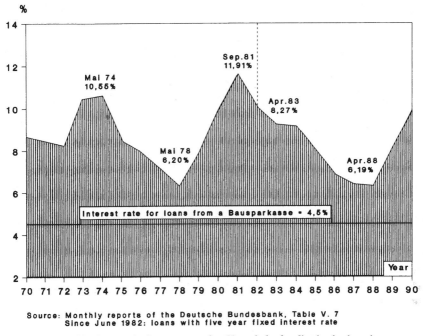

Figure 4.3 Effective yearly interest rate for *Hypothekarkredite* in the housing sector.

current interest rate in Germany, nearly 10%, is high and is a major factor hindering the development of new dwellings.

Repayment terms are generally flexible, but in the residential sector a repayment over 30 years at an annual rate of 1%, sometimes 2%, is usual. In the industrial sector, repayment rates are shorter and interest rates higher because industrial or office buildings are considered to have a shorter useful life.

Another important factor affecting the financing package is the gearing at which the banks are prepared to grant a mortgage or loan on a property (*Beleihungswert*). The principles for granting mortgages are fixed in guidelines for each of the above listed kinds of institutions and are based on the common rules of the *Hypothekenbankgesetz* (19 December 1990). The principles are similar although the details differ. For example, the Landesbank of Nordrhein–Westfalen (WestLB) has laid down the principle that the Beleihungswert should be based on market value, assessed along similar (but not identical) lines to those used by the valuation committee (Gutachterausschüsse), and a deduction of 10% to 40% should be made from this to reach the amount on which a mortgage will be granted (Weyers 1990: 77).

The deduction depends on the kind of project and its market sector, as well as the risk involved, as assessed by experts in the bank (e.g. low risk: a single housing project with deductions of up to 10% or an industrial project with deductions of up to 15% or high risk; single housing project with more than 30% deduction or an industrial project with more than 40% deduction).

However, by law a mortgage cannot be granted on 100% of the Beleihungswert: banks may only grant mortgages registered as a first charge in the land register at up to 60% of the Beleihungswert. A second mortgage registered at the second rank in the land register could be granted at up to 80% of the Beleihungswert, but higher interest rates would be payable.

For example, the mortgage given for a project with a market value of DM300,000 judged by a bank to be a medium risk (i.e. attracting a 20% deduction and therefore giving a Beleihungswert of DM240,000) would be a maximum of DM144,000 (60% of the Beleihungswert or 48% of the market value). A second mortgage at a higher interest rate could be obtained for a further DM48,000 (a further 20% of the Beleihungswert, equal to 16% of the market value). This leaves one-third of the market value of the property as a reserve margin for the bank's security.

The financial environment in each sector

The following financial assistance is available to building projects (see also Chs 4.3 and 8.3):

○ public subsidies given by Bund, Länder or municipality for special projects (e.g. low-rent housing, urban renewal, relocation of businesses, etc.);
○ public loans (low-interest or interest-free) given by Bund, Länder or municipality for special projects (e.g. low-rent housing, support for businesses, etc.);
○ tax advantages, especially in the case of new buildings;
○ other public subsidies (e.g. within the land policy: offering cheap plots, deducting servicing charges for plots, etc.);
○ self-development by the owner-occupier (there is a possible saving of 5–10% on construction costs in the owner-occupied housing sector);
○ equity capital (e.g. savings, shares, inheritance, etc.);
○ loans given by banks at market-orientated interest rates:
 (a) mortgage up to 60% of the Beleihungswert.
 (b) mortgage up to 80% of the Beleihungswert;
○ loans given by building societies at low, fixed interest rates;
○ loans given by an employer (low-interest or interest-free);
○ Ib-Hypotheken: loans (lagging behind the others) given by banks, etc., but only with public guarantees as security and;
○ Lastenzuschuß: special subsidy in the owner-occupied housing sector.

Financing the housing sector

The following remarks aim to clarify the financing of the housing sector, with reference to the above list:

○ If Bund and Länder grant direct subsidies for a project, this is called publicly promoted housing (*öffentlich geförderter Wohnungsbau*) (see Ch. 4.3). Other projects receive tax relief (*steuerbegünstigter Wohnungsbau*) or are part of the market-financed housing sector (*frei-finanzierter Wohnungsbau*).

○ The investment yields in the housing sector are fairly low, totalling 4% in the low-rent housing sector and about 5% in the market-orientated housing sector. Rents for new houses have increased markedly over the past two years (by 10% or more) and therefore returns have also increased, with resulting social problems.

○ Self-development reduces the necessary loan capital and is used in almost one-third of owner-occupier projects (mainly in one- or two-family houses). On average, 5–10% of the costs of a private contractor can be saved. Self-development is especially important in rural regions.

○ Equity capital may be savings, shares, savings in a building society, ownership of an unburdened plot, etc. Additionally, inherited or transferred wealth in money or property is part of the equity capital and will become more extensive in the near future. With regard to the housing sector in general, the individual builder should finance 30% (with a minimum 20%) of the total costs with equity capital. Actually in 1989 an owner-builder in work needed on average seven times the net annual salary to purchase a house or flat at an average value of DM280,000. The costs were financed with DM82,000 (29%) equity capital and DM198,000 (71%) loan capital (GEWOS 1990a: 11). According to the group of building society investors, on average 40% of costs are financed with equity capital, usually coming to DM117,000 in the case of new dwellings and DM81,000 for used dwellings. Typically, half these amounts are building-society savings and half savings in banks. In 15–20% of cases, equity capital is even higher because of inheritance, or in 6% of cases because of proceeds from the sale of a former house (Raschke 1991: 260). The proportion of equity capital is an important measure for banks lending the loan capital, as a healthy amount represents a base for the solidity of a project.

○ Loan capital is mainly borrowed from either Hypothekenbanken, savings banks and sometimes life insurance companies, or from building societies. The Hypothekarkredit, as described above, is the most important type of loan capital. On average 70% of building costs are financed by loan capital. Generally the share of loan capital borrowed from a building society is relatively small. A survey of its customers (owner-occupiers) taken in 1989 by the *Landesbausparkasse* Münster/Düsseldorf showed that after the 40% equity capital, only about 15% of costs were covered by building society loans, so the remaining 45% of costs must have been covered by bank loans.

○ Savings/borrowing contracts with building societies (*Bausparen*) is widely used in Germany, especially by owner-occupiers. It was introduced in 1931 to help financially weaker sections of the population to become owner-occupiers. The government subsidizes savings held in a building society by offering premium accounts or tax advantages (*Bausparprämien*) (see Ch. 4.3).

In a savings/borrowing contract with either a private or public building society, the amount of the contract is agreed upon (e.g. DM100,000). The rules stipulate that the customer should save 40% of this sum for a seven-year period; the investor is then entitled to a loan of 60% of the sum. In practice, more than 50% is often saved and loans are of the same order or even smaller (Statistisches Bundesamt 1990: 324). The main advantage of the system is that interest rates are low and at a fixed-rate (4.5–6.5%). Another advantage is that a mortgage of the third rank in the land register is accepted as a security. However, the interest rate paid on savings is also low (2.5–4.5%); the spread between the loan rate and the savings rate is usually 2%. And the burden is fairly high because the loan normally has to be paid off within 12 years at a monthly charge including interest of 1% of the loan (BMBau 1990c: 68). Building society customers often do not take full advantage of the cheap loan and prefer to use the contract as an instrument of saving.

The loan may be used for many purposes: purchase, construction, or extension, modernization or renovation of a house or flat, plot purchaser, modernization (as a tenant), paying off co-heirs, paying inheritance tax, or financing commercial rooms within a residential area or flats in old people's homes. Since 1991 it has been possible for building societies to be active in all EC member states.

In 80% of cases where property is bought for owner-occupation, building society savings were involved (Raschke 1991: 259; BMBau 1990c: 67). These figures demonstrate the importance of the Bausparen within the German housing sector. On average, savings and loans from a building society amount to one-third of the total costs of an owner-occupied housing project (Bau- und Wohnfibel: 67).

○ The conditions of a loan given by a Hypothekenbank or savings bank are described above. Loans given by life insurance companies are peculiar in that only interest, not the principal, is paid in the periodic repayments. The other component of these is a premium on a life insurance policy; the principal of the loan is paid out of the proceeds of the life assurance policy when it matures. Only about 10% of housing projects are financed in this way (GEWOS 1990a: 11).

Investors in the housing sector

Investors play an important rôle in the land and property market. They are described within the financial framework of the market, but also create demand for building land (see Ch. 5). Apart from owner-occupiers themselves, the following categories act as investors in the housing market:

○ Private persons (landlords) invest their money in housing projects with two different objectives: some of them want to diversify their investments (putting them into property, shares, savings, gold, etc.) in order to decrease risk. In this context, housing investments are considered stable but low-yielding. Furthermore there are strong legal controls on rent levels. Such investors normally have a high proportion of equity capital, but nowadays they often prefer to make their investments in property market sectors with higher returns (e.g. the commercial sector) or non-property investments.

The second group of private investors aims to create debts in order to reduce their tax burden. These investors act almost exclusively with loan capital as they are allowed to set the running costs, especially interest, against taxable income. Special patterns of investment have arisen, for example, the *Bauherren-, Bauträger-* or *Erwerbermodell*. However, these projects have failed too often and so have acquired a bad reputation. Similar objectives lie behind real-estate funds with a limited number of participants (*Geschlossene Immobilienfonds*), but they do not normally invest in the housing sector.

○ In addition to private builders, housing associations (*Wohnungsbaugesellschaften*) also invest. First, there are the companies acting under market conditions (*freie Wohnungsunternehmen*) and secondly, the non-profit-making societies (*gemeinnützige Wohnungsunternehmen*). However, the Gemeinnützige Wohnungsbaugesellschaft was deleted by law in 1990. Their former tax advantages and limited profits have been abandoned and they now have to act under market conditions. There is normally one former non-profit-making society in each town or district, generally closely connected to the municipality (see Ch. 7.1 and 7.2).

○ Co-operative societies (*Wohnungsbaugenossenschaften*) invest their members' money in housing projects and are socially orientated societies whose members have the right to live in one of the flats owned by the society. They are non-profit-making companies.

○ Another group of investors is categorized as building contractors (*Bauträger*), although they normally act as intermediaries (see Hildesheim and Stuttgart case studies, Ch. 7.1 and 7.2). Speculative developers are not very active in the housing sector. Other minor investors are the open-ended real-estate funds (*Offene Immobilienfonds*), which normally invest only a small proportion (less than 10%) of their property wealth in the housing sector. Insurance companies and pension funds are also rarely involved (the share of housing investments is less than 5% of their portfolio and often less than 1%).

Yields in the housing sector are not generally very attractive. Only in well situated locations and under favourable market conditions (e.g. after a period of inflation) do they reach the level of non-property capital investments. However the attractiveness of these investments depends on the stability and increase in the property's value (Falk 1985: 490).

Investors and financing in other market sectors

The office sector is currently the most profitable in the property market in all larger German towns. Returns on investments are normally more than 10%. Only in the top locations are the yields less.

In these markets the real estate-funds (*Immobilienfonds*) are the main investors, especially for large investments in or next to office locations in Frankfurt, Hamburg, Berlin, Düsseldorf or München (see Ch. 11.2). The open real-estate funds (Offene Immobilienfonds) are of great importance. They are managed by a capital investment company and each fund has a close relationship with a bank. They have to act within a legally fixed framework, the Act for Kapitalanlagegesellschaften (KAGG) 1970. For example, the risk of investment is limited and the fund is controlled by the Bundesaufsichtsamt für das Kreditwesen (federal office for credit matters). At least 65% of the fund's wealth must be in property. These investors are usually dealing with 100% equity capital (Falk 1985: 128). There are about 10 of these open-ended trust funds in Germany.

The closed real-estate funds (Geschlossene Immobilienfonds) invest in smaller-scale office projects. There is no specific legal framework regulating their action. The funds invest using large shares of loan capital, averaging 70% in 1984 (Falk 1985: 123). Tax advantages derived from interest payments, amortization and running costs are important to the shareholders. Shareholders are considered owners and take up tax advantages as landlords, as described above.

Insurance companies, including pension funds, are also investors in the office sector. However, as a percentage of their total wealth, their property assets have decreased since the 1970s. By the end of the 1980s only 7% of the total investments of life insurance companies and pension funds were in property, mainly commercial (Platz 1989: 41). However, their investment should increase again because of improved economic conditions and the repeal in 1987 of the law limiting investment in commercial property to a maximum 10%. The action of insurance companies is laid down in the *Versicherungsaufsichtsgesetz* (VAG) 1983 (legislation for controlling insurance companies) and controlled by the *Bundesaufsichtsamt für das Versicherungs-wesen*. The life insurance companies are the most important of the insurance groups, owning more than 60% (equivalent to DM25 billion in 1989) of total insurance company funds. The pension funds have a share of about 10% (nearly DM5 billion in 1989) of the total wealth (Statistisches Bundesamt 1990: 330).

Private persons also invest in the office market, particularly buildings with a mixture of offices and flats, and also in secondary business areas surrounding the city centre. Another important group are the companies constructing offices for their own use, for example, as company headquarters.

Finally, there are the foreign investors, who prefer to invest in the office and retail sectors (see Ch. 11.2). There are no restrictions on capital inflows, but investors must accept German legal rules. However, investment companies are not controlled by the Bundesaufsichtsamt, although they do have to meet certain conditions designed to protect shareholders (Model et al. 1991: 933). Companies usually organize their investments in such a way that the returns for foreign investors are tax-free.

Generally the conditions of investment, the investors and their behaviour are similar in the both the retail and the office sectors.

In Germany, the industrial sector has traditionally been dominated by investment undertaken by owner-occupier companies, especially in the business-park sector (*Gewerbeparks*) (see Ch. 11.1), although in recent years the number of development projects with rentable production space has increased. New investments are encouraged with attractive amortization rates for owner-occupiers as well as investments of landlords (see Ch. 4.3). In the investment market, investors are often open-ended or closed-ended real-estate funds as well as insurance companies.

Land banking

In Germany, land banking in respect of the *Bodenvorratspolitik* is mainly the task, although not a legal duty, of the municipalities. The level of involvement consequently varies greatly between different municipalities. Land banking is used to influence the land market in order to implement plans. The financial aspects of this instrument are very important but are nevertheless not the most significant part. This is discussed further elsewhere in this chapter (see Ch. 4.1 and 4.3) in the context of instruments for the implementation of plans and subsidies affecting the land-development process. It is also of importance in the case studies concerning the land market (Ch. 7).

No exact information about private land-banking is available, but it is not uncommon in private companies in all market sectors. Very often it is combined with speculation (see Ch. 6.7).

Transaction costs

The land market involves many different costs, fees, charges and commissions. Most have to be paid by the purchaser. Some of the costs arise when the ownership of a property changes, others when a site is developed.

The services of a real-estate agent do not necessarily have to be called upon, and they are still rarely used in some sectors of the market. The agent is only entitled to commission if a contract is drawn up for a property or rental flats. There is no legally fixed fee; it is generally negotiable for each case. Usual percentages are, however, published by agent's associations. Normally, 3% of the purchase price plus VAT has to be paid by both the seller and the buyer (Dehnen 1990: 15). In the most profitable market sectors (e.g. office buildings in major cities) fees of up to 5% of the purchase price are usual. In the rental sector, the agent's fee for setting up a rental contract is two to three months' rent (see also Ch. 7).

Changes of ownership only come into force when the entry in the land register (Grundbuch) is changed, and this is not possible without taking up the services of a notary. The fees are paid to the notary, who pays the local court (*Amtsgericht*). The fees are legally fixed (*Kostenordnung*) and are dependent on the purchase price and the legal contents of the document. In normal cases the fee is equal to 1.5–2% of the purchase price (plus VAT) (Dehnen 1990: 16).

It is generally agreed that the buyer pays the notary's fees. In most cases of registering a change of property owner, the notary and the local court are consulted a second time if a mortgage has to be registered in the land register. The rules for the fee, which is normally 0.5% of the purchase price, are also to be found in the Kostenordnung.

The land transfer tax of 2% of the purchase price has to be paid by the seller or buyer; it is usually agreed that the buyer pays. The commission of a real-estate agent is also subject to land transfer tax (for details see Ch. 4.3).

These fees, which are normally unavoidable, come to about 7% of the purchase price, and can be deducted from taxable income. In nearly all cases, the cost of insuring a property must be added, although, except for fire insurance, this is not compulsory. It is usual to insure against storm and water damage and building liability. The banks insist on insurance being taken out before giving loans.

When a new owner buys an unbuilt plot and intends to construct a building, the following costs occur in addition to transaction costs:

○ Charges for the services of local authorities vary, but are fixed by law in each of the Länder. For example, a fee of up to DM500 is charged for permission to divide a site into smaller plots. The costs of surveying a site in order to create smaller, regular plots are fixed by the Länder in a scale of charges specified for surveying (the VermGebO of 1973). The costs depend on the land value and size of the plot and equal, on average, DM5-7 per m^2 of building land. Planning permission costs 0.5-1% of the estimated construction costs.

○ The fees paid for the services of architects and engineers are fixed by law in the HOAI (a special scale of charges for architects), and increase in line with the construction costs of the building and the complexity of the project. They come to 8-15% of the construction costs plus VAT; the percentage is normally lower for residential buildings than for commercial buildings.

○ When a building is finished it has to be measured for the cadastre. These fees, paid by the owner, depend on the value of the building and are normally 0.1-0.2% of that value.

○ The owner has to pay 90% of the charges for servicing a plot with reference to the Erschließungsbeiträge (for details, see Ch. 4.1).

○ The owner has to pay the costs of connecting a property to the mains services (e.g. water, sewerage, gas, electricity, telephone).

The additional costs and fees of constructing a new building are equal to about 10-15% (and can often be up to 20%) of the calculated construction costs, and development fees have to be added to this. These costs are classed as *Baunebenkosten* (additional construction costs) and normally come to 15-18% (Simon et al. 1991: 601). They form part of the total construction costs that are tax-deductible.

4.3 The tax and subsidy environment

Property-related taxes in Germany are generally based on the total value of the land and the building elements, rather than each of the components. Therefore no differentiation can be made between the land and property markets.

Overview of German property taxation

Taxes on land and property include:
○ land and buildings;
○ changes in ownership;
○ increases in capital value and;
○ total return.

The following taxes are important in the German land and property market:
(a) property tax (*Grundsteuer*);
(b) wealth tax (*Vermögensteuer*);

(c) death/gift duties (*Erbschaft-/Schenkungsteuer*);

(d) land transfer tax (*Grunderwerbsteuer*);

(e) income tax (*Einkommensteuer*);

(f) speculation tax (*Spekulationsteuer*);

(g) corporation tax (*Körperschaftsteuer*);

(h) local business tax (*Gewerbesteuer*); and

(i) value-added tax (*Mehrwertsteuer*).

These can be grouped as follows:

○ taxation of property itself (a), (b);

○ taxation of changes in ownership (c), (d), (f); and

○ taxation of returns from property (e), (g), (h).

Figure 4.4 gives an overview of total tax revenues in Germany, with the revenues summarized for the three tiers of the federal state.

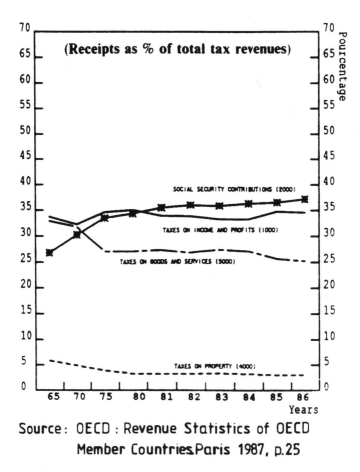

Source: OECD : Revenue Statistics of OECD Member Countries Paris 1987, p.25

Figure 4.4 Tax revenue in Germany.

The first three taxes (a–c) are derived from the same base, the standard value of property (Einheitswert – EHW) which is fixed by the financial administration. Their legal base lies in the *Bewertungsgesetz* (BewG) 1991 (legislation for valuing properties). Valuation methods are based on cost or total return. A problem exists in the valuation method in that it still refers to 1 January 1964, despite the fact that the Bewertungsgesetz demands a new valuation every six years. Since the mid-1970s an extra charge of 40% has been added to the values, so today the standard values are the real values of 1964 plus 40%. The difference between this standard value and actual market value has become larger and larger because of greatly increased land and property prices, including inflation (see Ch. 1.4). Today they are on average about 10–15 % of real market values (Haasis 1987: 63).

Detailed information about the different taxes relevant to the land and property market in Germany is given in Table 4.2, and the following section should be read in conjunction with it.

Some remarks about each tax and its importance
Property tax The amount of the property tax is based on the (very low) EHW (standard value) and is dependent on the Steuermeßzahl (SMZ – tax measure number) and the Hebesatz (HbS – tax increase number). The Steuermeßzahl is fixed by federal law as follows:

Normal cases	0.35%
Single-family houses	
– up to DM75,000 EHW	0.26%
– more than DM75,000 EHW	0.35%
Two-family houses	0.31%
Agricultural/forestry land	0.60%

The Hebesatz is fixed by each municipality. In 1988, the rate varied between 250% and 440%, averaging 303% (Statistisches Bundesamt 1990: 478). In general, the Hebesatz multiplier increases with the size of town. In most of the municipalities there is a special Hebesatz-rate for agricultural/forestry land, usually lower than the basic rate. A third, usually higher, rate is applied to industrial and commercial business, in connection with the local business tax (Gewerbesteuer).

The property tax is not a large burden on owners. It is roughly estimated to burden the taxpayer with 0.1% of the open market value each year and brings the municipality about 10% of its revenue.

In 1988 and 1989, total tax revenue contributed only 40% of the total revenue of a municipality. The rest comes from contributions from the federal government (Bund) and the Länder (22%), charges for the supply of drinking water, waste water and refuse disposal, street-cleaning and administration fees (25%) and others (13%) (BMBau 1990a: 121).

Table 4.2 Relevant land and property taxes in Germany.

Tax	Taxable person	Object of the tax	Base of the tax	Amount of tax in %	Tax exemptions	Recipient	Proportion of the total tax revenue of the recipient 1988	1989
Local property tax	owner, persons owning property by hereditary right except: public institutions, and the church	–all kinds of land and buildings (incl. agricultural land) –condominiums –hereditary rights	standard value Einheitswert EHW HbS 1988 220–440 av = 300	1% x EHW; 0.1% x market value	today nil	Municipality	12	11.5
Wealth tax Vermögensteuer	natural persons or households, and corporations	total wealth including land and property	wealth in DM land and property based on Einheitswert	Tax = 0.5 x private wealth, 0.6 x for corporations	all debts can be deducted in market values	Länder	1.2	3.0
Death/ Gift Duties Erbschaft/ Schenkungsteuer	heir or recipient of gift	total inheritance/ gift also permission to occupy flats without rent	Land and property: standard value. Rest market value.	Depends on the amount received and the family relationship	Care allowance plus other family allowence	Länder	1.4	1.1
Land transfer tax Grunderwerbsteuer	seller and buyer see contract	change of ownership including inheritance	value of the contract including fees etc	2 % of the value of the contract	Transfers between marriage partners and near family, low purchase price	Länder (90%) Municipality (10%)	1.7 0.4	1.9 0.4

Table 4.2 (continued)

Income tax Einkommensteuer	natural persons	income	income in seven types -agriculture -capital -commercial enterprises -self employment -renting	<5,000 = tax free 5,000–8,150 = 19% 8,150–120,000 = 19–53% progressive >120,000 = 53%	many different exemptions possible	Bund (42.5%) Länder (42.5%) Municipality (15%)	47 60.5 44	46.6 61.5 44.5
Speculation tax Spekulationsteuer part of income tax	seller, former owner	land, buildings sold within two years of purchase	profits from sales	profit is regarded as income and taxed accordingly	profits <1,000DM are tax free	cf income tax	cf income tax	
Corporation tax Körperschaftsteuer	al corporations	income	taxable income	normal rate = 50% since 1990	for special corporations, eg pension funds	Bund (50%) Länder (50%)	cf income tax	
Local business tax Gewerbesteuer	commercial enterprise -with exceptions	returns or capital of business	returns per year + the EHW	Hbs 1988 250–400 average equalled 360	various	Bund (7.5%) Länder (7.5%) Municipality (85%)	1.1 1.4 42.8	1.0 1.4 42.6

Wealth tax (net worth tax) The base for the wealth tax is the total wealth (not only land and property wealth) of a taxpayer. Wealth in land and property is, however, favourably taxed in comparison to other forms of wealth because it is assessed with the obsolete and low standard value.

The tax allowances are high, for example DM280,000 for a family with two children. Because of the low standard value this would correspond to a wealth in land and property of DM2 million. Therefore a household in Germany owning a single- or a two-family house does not normally pay wealth tax on its property.

In addition, debts are completely tax-deductible. If, for example, a property investment of DM2 million is made (with a standard value of DM280,000 as above) and only 25% is financed by loans (i.e. debts of DM500,000), the owner does not have to pay any wealth tax for this property and, indeed, he may set the balance of DM200,000 off against the rest of the taxable wealth.

It may be concluded that wealth tax is not paid by many people in Germany (only about 700,000 people in 1986). It contributes about 3% to the revenue of the Länder.

Death/gift duty The amount of death/gift duty depends on the family relationship between the testator/donor and the beneficiary. If the family relationship is very close (e.g. marriage partner or children) the tax is fairly low and high allowances are taken into account. On the other hand the tax can be very high (e.g. up to 50% and more on the heir) if the beneficiaries are only distantly related and if the amount inherited is large.

The inheritance of land and property also benefits from the low evaluation of standard values (based on 1964) plus 40%. The examples given for the wealth tax are also applicable to death/gift duties.

When inheriting, an heir need not pay land transfer tax, income tax or VAT.

Land transfer tax Until 1982, the land transfer tax was more important, as then it amounted to 7% of the purchase price. There were, however, a confusing number of exemptions and the rules differed slightly in each Land. Since 1983 there has only been one legally binding federal statute fixing the tax at 2% of the purchase price. Nowadays only a few insignificant exemptions remain. The base of the tax normally corresponds to the purchase price and it generally includes all the outgoings the purchaser incurs in acquiring ownership.

Purchasing a new property often involves two steps: a contract for the unbuilt plot and a second contract for the construction of a building by a builder. In such cases the land transfer tax is based only on the price of the land. However, if the only reason for concluding two contracts is to avoid tax, for example, buying an already completed dwelling from a construction firm with separate contracts for land and building, the authorities take this into account.

The entry in the land register (Grundbuch) may only be changed if the tax authority (*Finanzamt*) confirms that the land transfer tax has been paid (*Unbedenklichkeitsbescheinigung*). The land transfer tax is now rarely avoidable, fairly low and generally accepted. It is therefore taken account of as a fixed cost. The Länder are the main beneficiaries of the tax, although it is a minor part of their revenue.

Income tax The most important tax for land and property owners, as well as property professionals, is income tax. Annual income less expenses, allowances and exemptions is subject to taxation.

Seven kinds of income are taxed. The most important categories for those in the land and property market are:

○ income from commercial companies (e.g. construction firms, real-estate agents, property developers);

○ income from self-employed businesses, e.g. architects and notaries (*freie Berufe*);

○ income from renting or leasing (e.g. the private owners). This category of income tax is the most important for the property market, but it only contributed a 1.5% share to total income tax revenue in 1983. Owner-occupation of a flat or house is not taxed.

Income tax is the most important source of revenue for the government (see Table 4.2). The Bund and the Länder each get 42.5% of the total revenue, while the municipalities get 15%. In the 1988 and 1989 tax years, income tax contributed 45% towards the revenue of the Bund and the municipalities and a high contribution of 60% towards the revenue of the Länder.

Income tax has a burdensome effect, but in practice there are many ways for property owners to use income tax exemptions to maximum benefit. For example, cost depreciation in the first years of ownership (see below) normally substantially reduces the negative taxable income in the category "income from renting and leasing". This loss can be set against other sources of income, resulting in overall tax advantages. For many people, house ownership is possible only because of tax reliefs. Such depreciation is one of the most important and effective instruments of the policy directing private money into the housing market.

Speculation tax Speculation tax is a capital tax grouped in the income tax schedules. In the property market, the target of the tax is profit from the purchase and sale of the same property within two years (speculation business). When the property has been owned for more than two years, profits are tax-free. Losses may be set against profits from other speculative business during the same year. The sale of an inherited/given property within two years is not subject to speculation tax.

The period used to decide whether speculation tax is payable is calculated from the date the deed is sealed by a notary, not from the date of the change in ownership. It is the agreements concerning the plot, not the building, that are decisive. The building counts towards purchase profit only if it was bought with the land, not if it was constructed by the new owner during the two-year period. It may therefore be necessary to split the total purchase price between land and buildings.

In practice the speculation tax does little to hinder speculation, because two years is too short a period in comparison to the time necessary to develop a plot or construct a building. It is economically beneficial to wait a few extra months, especially in times of rising property prices.

Corporation tax Corporation tax (Körperschaftsteuer) is very similar to income tax, but an important difference is that there is no progressive tax rate. The rate is fixed at 50%, reduced to 46% in special cases. These rates were higher before 1990 (56% and 50%, respectively). The net worth in property is calculated using the low standard value (Einheitswert).

There are many possible total exemptions for specific types of corporation, for example, pension funds whose income does not exceed special limits set depending on their net worth. Insurance companies can also deduct large reserves from taxable income and reduce corporation tax.

Local business tax The local business tax (Gewerbesteuer) is composed of two parts. One taxes the return of the enterprise/business (*Gewerbeertragsteuer*) and the other taxes the capital (*Gewerbekapitalsteuer*). Both parts take into account the fact that land and buildings have already been taxed by the property tax. There are reductions orientated towards the standard value; double taxation should be avoided.

Tax on business return is much more important than tax on capital. It comes, on average, to 15–20% of the total return (return is defined and calculated in the same way as for income tax), while the tax on a business's capital comes to only 0.5–1% of the standard value of all the capital (less land and buildings). In addition, the allowance for capital is more than three times the allowance for return.

In the case of property, this tax is of less importance to company owners. However, if an owner intends to sell or abandon the company, the proceeds of the sale, including that of the land, are subject to taxation as return at an average rate of 15–20%. Furthermore, the local business tax is important for all people earning a salary by dealing in land and property or offering services in this field (property professionals).

It should be noted that, if a person deals in property for economic reasons, such as renting out privately owned flats, but not for reasons of business, it is not normally considered to be commercial activity and so is exempt from local business tax. These persons have to pay income tax on their earnings from renting and leasing. However, if a person's main activity is the renting of flats, that person is regarded as being in business and he becomes liable to the local business tax (and income tax is paid as for commercial or industrial activities). The courts have ruled that if someone develops a site with the intention of constructing and selling up to four flats, this is still considered administration of private wealth.

Besides income tax, local business tax is the most important source of revenue for the municipalities. Although in terms of the property market the tax is of less importance, it does affect the land market and especially that for industrial and commercial land. The importance of local business tax in their revenue causes the municipalities to be financially dependent on the prosperity of companies within their boundaries, and this can lead to competition between municipalities. In accordance with the proverb "if two are quarrelling, the third is pleased" the municipalities often get themselves into disadvantageous positions when negotiating with companies. This is a major reason why the industrial land market functions differently from other market sectors and is characterized by low prices and many "secret subventions". Only the most powerful municipalities in Germany (for example, München, Stuttgart, Frankfurt and, no doubt in the future, Berlin) are able to avoid this dependence.

Each municipality has to fix its *Hebesatz* (HbS) and so is itself responsible for the level of local business tax. This rate is an important criterion for the location of a company. If a company produces in different municipalities, each of these receives a proportional share in taxation of the relevant returns and capital of the company.

Value-added tax In principle, land and property deals are not subject to VAT. Income from renting and leasing is also free from VAT. In Germany the current VAT rate is 14%, which the government intends to raise to 15–16% to achieve harmonization within the EC.

There are possibilities for companies, although not private persons, to save tax. A company may credit the VAT it pays for developments (*Vorsteuerabzug*) (e.g. construction of a building), if it voluntarily pays tax on the yearly income from renting the building. This possibility may bring tax advantages, particularly in the construction phase. It is said that this would finance 9% of all construction costs. However, since 1985 this procedure has not been permitted for housing projects, although it is still possible in the industrial and office sector as well as for the renting of vacation houses, studio flat hotels or flats for NATO.

Subsidies for the land-development process

In Germany the land development process consists of both land-use planning and the creation of conditions enabling private owners or participants to implement the intended use.

The first part of the process, land-use planning, is the duty and right of the municipalities (see Ch. 1.1). The second part is not their exclusive right, but in practice it is also their job, as they have to service the plots. So direct subsidies concerning the land development process would in most cases favour the municipalities.

Planning costs usually have to be carried by the municipality. The costs of changing plots or boundaries to prepare a plot for building have to be carried by the landowners. If structures of ownership present difficulties, there are legal instruments for changing boundaries, for example, the replotting procedure (see Ch. 4.1). The administrative costs have to be carried by the municipality and no subsidies are available for this. Before 1983, the equivalent of a subsidy existed for voluntary, privately initiated replotting procedures. Within this procedure landowners had been free of land transfer tax. This has been changed in connection with the lower 2% rate of the land transfer tax (Dieterich 1990a: 318).

Subsidies for the cost of installing public utilities (technical infrastructure) are also not usually necessary, as the municipalities can split 90% of the costs between landowners who benefit (see Ch. 4.1). The remaining 10% of the costs paid by the municipality is not subsidized by the higher tiers of government. However, this part of the costs is not considered as a subsidy for the landowners, but the price of the advantages to the municipality brought about by the development, and an incentive for realistic planning and development.

However, for the development of industrial land in particular, the municipalities often carry more than 10% of development costs in order to stimulate companies and increase their own competitiveness with other municipalities. The extent of these "secret subventions" is unknown. Grants from the *Städtebauförderungsmittel* programme (grant for urban improvement) are often used by the municipalities to service industrial building land.

Nevertheless there are three main types of subsidies for special items which affect the land-development process:

Subsidies to promote the development of the economy The German economy is to a large extent promoted by the many subvention programmes given by the federation and the Länder, to help both entrepreneur and municipalities.

94

Among their aims are the promotion of new business (*Existenzgründer*), new technology, small and medium-size firms, and environmental protection or energy efficiency in industrial production. The subsidies are awarded in different forms by the Bund and Länder (grants, bonus payment, loans, guarantees). Once a year, all relevant and available subsidies are listed in a special edition of the magazine *Zeitschrift für das gesamte Kreditwesen*. Further issues of this magazine list the subsidies available within the residential sector (No. 2) and the agricultural sector (No. 3). It gives an excellent overview of the possibilities for specific purposes and the conditions and means of application, but unfortunately does not quantify the grants. Figure 4.5 includes the above-mentioned programmes within the unspecified DM13.5 billion of subsidies for "business and industry".

The *Gemeinschaftsaufgabe Verbesserung der regionalen Wirtschaftsstruktur* (improvement of the structure of the regional economy) is an important programme for which Bund and Länder are jointly responsible and finance equally. It aims to reduce regional disparities and include regions in the north and the old-industrialized zones (see Fig. 4.5 under "Business and Industry" and "Subsidies because of the two Germanys"). These regions have 40% of the west German population. The costs of special investments can be subsidized by between 10% and 23%. The costs for industrial land or buildings are included in the granted investment costs. In 1989 the programme granted DM1 billion, of which private investments were granted two-thirds (mostly for enlargements of firms) and infrastructure investments one-third (BMBau 1990a: 106). It is to be expected that in the next few years these subsidies will be concentrated on the new Länder in eastern Germany.

It is not possible to quantify the influence of these subsidies on the demand for industrial land or on its price. It should be noted, however, that the programme only covers rural and old-industrialized areas where the land and property markets are not booming. This is also the case for the few regions for which EC funds are available.

The promotion of economic development is also pursued by each municipality (*Kommunale Wirtschaftsförderung*) and the chambers of commerce. The importance of companies and the tax they pay to the financial standing of municipalities, and the resulting competition between them was stressed earlier.

The municipalities are often the intermediate owners of the land they intend to develop for industrial use (see Ch. 7.3). This allows them to offer reduced price plots to new businesses. Another common way of subsidizing a company is to reduce the development fees/servicing costs or to promote its resettlement within the municipality (Heuer 1985). However the budget problems are the same as mentioned before.

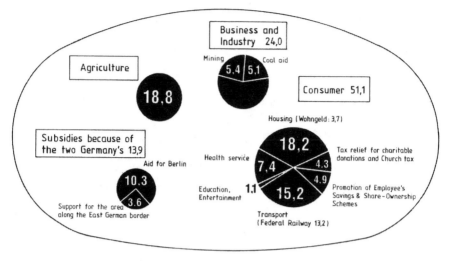

Source: Westfälische Rundschau vom 9.1.1991

Figure 4.5 Yearly subsidies classified in four categories.

Redevelopment of contaminated land The redevelopment of contaminated land is an important issue in many of the larger German towns, especially in the older industrialized zones. These areas normally have good access to road and rail networks and so are well suited for redevelopment. The owners, however, shrink from the costs of cleaning up contamination although, in principle, the person who caused the damage should bear the costs of eliminating it. Only in the south German agglomerations are prices for industrial land sufficiently high to warrant the financing of decontamination.

Subsidies for cleaning and re-using derelict land within a fixed redevelopment area (§136ff. BauGB) can be granted by the Städtebauförderungsmittel. These subsidies are given by the Bund, Länder and municipalities together in a common programme for urban improvement and redevelopment (see below). Other possibilities for subvention are given within the Gemeinschaftsaufgabe Verbesserung der regionalen Wirtschaftsstruktur programme referred to above.

In addition some Länder issue their own programmes (land funds) for the redevelopment of contaminated land, often using the instrument of intermediate land ownership by the Länder or municipality. Particularly noteworthy was the foundation of the *Grundstücksfonds* in Nordrhein–Westfalen.

Providing of restricted-price building land Municipalities are legally obliged to provide enough cheap land for housing, especially for low-income groups

(§89(2) II.Wohnungsbaugesetz 1990), so the provision of building land at a reduced price is an important part of their land policy.

Different local land policies are applied, such as long-term land banking and short-term intermediate ownership by the municipalities. A municipality can sell the plots either at a reduced price orientated towards its own costs or at a fixed price agreed upon by the city council. This kind of subsidy is often used for land for special groups with social welfare objectives (see Ch. 7.2). The prices of the land are reduced by 50% or more of the market prices (BMBau 1986: 128).

The sale of reduced-price land is more common in the industrial land sector because of competition between municipalities, as explained before. An obstacle, other than the financing of the land itself, that exists is the budget legislation that requires a municipality to handle its funds economically. Nevertheless, such subsidies are common.

The Bund and Länder, as well as the municipalities, are asked to sell land from their stock at a reduced price; this land is particularly required for the low-rent housing programme. There is currently much discussion about the land and barracks left derelict by the reduction of armed forces. The same problems mentioned above exist in this case. The Bund wants to sell land to the municipalities for low-rent housing with a reduction of only 15% of the market price (Thielges 1991: 29), but the Land Baden-Württemberg reduces the land price by up to 80% (Finanzminister Baden-Württemberg 1990).

Because of current problems in the housing sector, the municipalities are able to grant cheap credit from the Kreditanstalt für Wiederaufbau for the planning and development of new housing land. The interest rates are 4% less than market ones and two-thirds of the costs could be granted.

Another subvention is given to farmers selling agricultural land for residential use. The requirement that these gains be re-invested in agriculture within two years is extended to four years.

Property subsidies affecting the land market

Tax advantages (indirect subsidies) In Germany, tax concessions for payers of income tax are very important to the government as instruments for influencing and regulating investment conditions in the land and property market. In this way, particularly in the housing sector, the state participates indirectly in the financing of each building. It is not possible, for example, to finance a semi-detached house in a profitable way unless tax advantages are taken into account. If they were not, monthly rent for living space based on actual costs would be much higher than the current market rents.

Tax concessions, especially various types of allowance for depreciation, reduce the tax liability of a taxpayer. Because of the progressive income tax rate, the reductions are more lucrative for higher total incomes. This socially unbalanced effect does little to promote the rate of ownership (owner-occupation). In addition, the possibility of crediting losses or transferring tax liabilities to subsequent years independently of income from other sources is very important for the taxpayer and has a large indirect effect on the decisions of dealers in the land and property market (Haasis 1987: 69).

There are many rules about tax relief in Germany. Some common cases are described here; it is not possible to give a complete survey. In recent years, tax relief has amounted to more than DM6 billion a year (BMBau 1990b: 53; see also Fig. 4.5 under "housing").

Table 4.3 Actual rates of amortization.

	Linear amortization	Degressive amortization years after finishing construction	
	%		%
A			
Non-resident buildings,		1–4	10.0
constructed since 1985,	4	5–7	5.0
part of firm's capital		8–25	2.0
B			
All other buildings constructed since 1925, e.g.			
all residential building, if not owner-occupied	2		
non-residential		1–8	5.0
building, constructed	2	9–14	2.5
1925–84		15–50	1.25
All other buildings, constructed before 1924	2.5		
C			
		1–4	7.0
Residential buildings		5–10	5.0
constructed since 1989,		11–16	2.0
if not owner-occupied		17–50	1.25

Source: Einkommensteuergesetz vom 7.9.1990, BGBE.I S.1898.

There are different rules for owner-occupiers and landlords. Once a building is finished and used, an owner-occupier may only increase his tax allowances by a fixed amount, while a landlord may also set against income interest payments, the costs of acquiring money, property tax, all other public contributions, building-insurance premiums and running costs (e.g. sewerage, refuse collection, and the services of a chimney-sweep). These regulations apply after the building is finished. During the relatively short period when a building is being planned and constructed, both owner-occupiers and landlords may deduct from their tax liability almost all the costs of the building process (e.g. costs for managing the construction and interest, but not the costs of the land and construction itself, the land transfer tax and charges for notaries and agents (§10e[6] EStG)).

There are different regulations for the amortization rate; the periods of amortization are, for non-residential company buildings, 25 years and, for residential buildings, in normal cases 50 years; but, when constructed by landlords since March 1989, 40 years

The *Einkommensteuergesetz* (EStG) differs between the following types of amortization and tax advantages:

(a) Linear amortization is the normal type. It may not be applied by owner-occupiers, but only by landlord investors for both new and used buildings.

(b) Degressive amortization can be used as an alternative to linear amortization and it brings more tax reduction in the first years. It can be applied only for new buildings. Landlords may use it for residential and non-residential buildings and owner-occupiers only for non-residential buildings. Degressive amortization is an important instrument in boosting the housing sector. Rates of amortization have changed in line with changes in government policy in 1981, 1985 and 1989. The rate depends on the year of finishing the building (new construction is to be encouraged).

As an example of amortization of a commercial building, take the case of the headquarters of an insurance company built in 1990 (owner-occupied office space). There is a choice between two amortization methods: 4% linear amortization for 25 years or degressive amortization starting with 10% and dropping to 2.5% from the eighth year (1998).

If the company built the office building as an investment project, the same amortization rates would apply. If a building was constructed in 1980 and bought in 1990, only the linear amortization of 2% would be applicable for the next 40 years (50 less 10 years of age). The same rates are applicable for owner-occupied industrial buildings.

Degressive amortization for new, non-residential buildings (office blocks, factory buildings, etc.) is rather high for the first years. This rate was introduced in 1985 to promote economic development. It is still applicable today ,although the economy has been booming for some years.

The promotion of owner-occupied residential buildings (single houses, flats in multiple dwellings and condominiums) is subsidized by means of tax relief. For eight years the owner gets tax relief, either in the form of a yearly 5% reduction of taxable income in respect of the total construction costs or through an allowance against the purchase price up to a maximum of DM300,000. The latter may include a maximum of 50% of the cost of the land (including all servicing costs). The maximum reduction allowed is DM15,000 a year. This is therefore equivalent to 40% of DM300,000 (DM120,000) deductible over eight years. Further amortization in the following years is not allowed. This incentive applies to new dwellings as well as to the purchase of used dwellings (flats or houses). Each owner may take advantage of this regulation once only, couples twice. Further improvements of these conditions have been discussed and became valid in 1992.

Another tax-relief vehicle for promoting owner-occupied dwellings is the *Baukindergeld* (child benefit when building a house). DM1,000 per child can be deducted from tax liability for eight years (§34f EStG). The Baukindergeld is dependent on the number of children up to the age of 16, or 27 if they are undertaking professional training (e.g. as students). The rate of the Baukindergeld is not dependent on the level of income.

An important change in the promotion of owner-occupied dwellings occurred in 1987. Before this these had been considered an investment that might be amortized: the owner could set the interest against income for tax purposes, but on the other hand had to count 1.4% of the standard value of the house as additional income. Owner-occupied flats are now considered to be consumer items not subject to amortization, so interest cannot be set against income tax; but neither is there any deemed additional income. Interest paid in their purchase is also not allowable against tax but they are not regarded as a source of taxable additional income. Whether the owner's position is improved by these changes depends on the particular personal circumstances.

In order to provide an illustration, the following rough calculation gives the example of a residential building. A family with three children is building a single-family house, assuming the following calculations:

Construction costs:	DM280,000
Costs of the plot:	DM100,000
Annual household income:	DM70,000

Corresponding income tax rate: 25%

Owner-occupied building: Tax reliefs:

25% of the maximum annual reduction of DM15,000: DM 3,750

Annual Baukindergeld for three children: DM3,000

The tax relief for eight years adds up to DM54,000

Where tax relief for a building with rented flats is concerned, linear amortization is rather low and would normally not be used by the investors. It comes to DM1,400 per year and to DM11,200 for eight years.

Degressive amortization (for buildings as from 1989) will come on average to DM4,200 per year for the first eight years and to a total advantage of DM33,600 over this period. In the following eight years the average yearly tax relief will be DM1,925. These sums are lower than those for owner-occupied property, but when account is taken of the allowance for depreciation, of the total yearly interest in particular, the landlord will normally be better off. On the other hand, taxable income will increase because of rental income.

(d) A special "raised rate of amortization", valid for new buildings constructed in the years 1989–92 is applicable only for landlords who construct flats whose rents are fixed at a relatively low level for 10 years. This is an alternative to the direct subventions of low-rent-housing (see below). Within the 10-year period, the costs of the building can be amortized by up to 85% (§7k EStG).

Direct subsidies

(a) The first category of direct subsidies, programmes for urban improvement (*Städtebauförderung*) involve all three levels of the federal state. Three types of support can be distinguished:

○ First, there is the promotion of urban change and improvement within a redevelopment area (*Sanierungsgebiet*). The area is fixed by local law and the redevelopment procedure is prescribed by §136ff BauBG (see Ch. 8.1). Subsidies are given to the municipalities as well as to house-owners or company-owners for improvements, for example, to housing stock, traffic situation, green areas or industrial redevelopment.

The improvement measures, which are a public duty, are financed by the Bund, Länder and municipalities together, each covering one-third of the costs. The Bund has given an annual DM660 million over the past three years, indicating that the total programme comes to DM2 billion (see Fig. 4.5, category "housing").

○ Each of the Länder also has its own programme promoting urban renewal projects. For example, Nordrhein–Westfalen gives grants mainly to municipalities, but also to private persons. Projects in the field of *Verkehrsberuhigung* (traffic reduction, or calming), social infrastructure, green space and especially the redevelopment of unused or derelict land, including the purchase of the unused land, are promoted. The percentage of state subvention is not fixed; it is often more than 50%, sometimes even up to 80%. The guidelines do not differ very much between the Länder.

○ The municipalities often set up their own programmes and grants for urban renewal and for improvement projects, depending on their financial strength.

(b) The contribution made by building society savings to the financing of housing projects was discussed in Ch. 4.2. In keeping with their aim of increasing owner-occupation, both Bund and Länder promote saving with a building society by giving building society investors (*Bausparer*) a premium account when the taxable income of the household is below DM54,000 (or half this sum for a single household). The subsidy (*Bausparprämie*) amounts to 10% of the yearly savings, up to a maximum DM1,600 (again half for singles). The Bund grants another subsidy, aimed at the same limited income group, separately from the building societies. Each employed person gets a premium of 10% for a yearly saving of a maximum of DM936 (*Vermögenswirksame Leistungen*) if the money is used for a different type of wealth creation, such as saving with a building society.

Another concession (*Sonderausgabenabzug*), which has no set limits on income, allows taxable income to be reduced by 50% of the savings with a building society.

In recent years, there has been less government promotion of saving with a building society. The state has gradually withdrawn from this type of promotion for lower-income groups because the government has continuously tightened the conditions and decreased the rates of premiums. The ratio between this kind of promotion and that springing from tax concessions changed from 1:0.6 in 1975 to 1:7.1 in 1990. In recent years the subsidy has totalled DM800–850 million.

(c) The subsidies of the Bund, Länder and municipalities within the social-housing sector aim to support the construction of new dwellings for low-income groups. However, in contrast to the Bausparen and the Lastenzuschuß there is no automatic entitlement to these grants. Social-housing grants are possible for rented flats as well as for owner-occupied dwellings (houses and condominiums). The construction of these properties is promoted in three different ways. Owner-occupiers receive grants if their income is below specified limits, and landlords receive grants when they agree to restrict rents and house special tenant groups.

In 1988 two-thirds of the financial aid from Bund and Länder was granted to private households (most of them for owner-occupation), 20% to non-profit building societies and 15% to other companies (BMBau 1990b: 48).

In 1987 the share of publicly supported flats as a percentage of all such accommodation averaged 15.5% overall and was 30% in the agglomerations (Raumordnungsbericht 1990: 90). The importance of low-rent housing declined until 1988. In 1978, 36.5% of all new flats were still constructed in the public house-building sector, but this percentage fell by half to 18.5% within 10 years (Haus und Wohnung 1990: 33, 45). In particular, the construction of flats for rent declined. During this period, the financial contribution of the Bund was reduced from 35% to 12% of the total grant; the rest of the grant being paid by the Länder.

Since 1988, because of real shortages in the housing sector, the efforts of the Bund and Länder have been greatly increased, especially for rented flats. In 1990 the grants were of a magnitude similar to that of the early 1980s (DM8.3 billion), with the Bund paying DM2 billion (24%) and the Länder DM6.3 billion. In 1990, 106,000 flats received grants (Bundestags-Drucksache 11/8158 in GuG 1/1991: 49). Statistics for the additional efforts of municipalities are not available.

The Länder are responsible for awarding grants, but the details of the process differ between them. The main legal framework is given in the II. *Wohnungsbaugesetz*, 1990. There are three kinds of support.

A first type of support (*Förderweg*) involves interest-free public loans (e.g. DM100,000), additional loans for families, or support to reduce the monthly running and mortgage costs (e.g. DM5.00 per m^2 of living space per month). A flat receiving a grant may not be larger than $130\,m^2$, a house with two flats no larger than $200\,m^2$, and a condominium no larger than $120\,m^2$. The limits on taxable income are less than for the aforementioned Bausparer premium accounts:
○ household with one person DM21,600;
○ household with four person DM47,800;
○ extra relief is available for young couples, handicapped persons, German emigrants from eastern Europe or investments in Berlin.

A second type of Förderweg includes loans that reduce the monthly payment (e.g. a monthly DM5.00 per m^2 of living space). The income limit is 40% higher than for the first type of Förderweg, the living space covered by the grant is normally a $90\,m^2$ flat, with a maximum area of 20% over this limit. The loans are for up to 16 years free of interest and repayment. It is assumed that by then repayment of a loan from a building society will be completed.

A third type of Förderweg (known as "arranged support") is a new instrument that should be more flexible than the others. The Länder or municipality and the investor agree upon the conditions of the support in a contract. The rents should be fixed for a minimum of seven years.

(d) If an owner-occupier has taken full advantage of the above opportunities for support and his remaining monthly commitments still exceed specified amounts (e.g. more than DM640 in the cheapest regions and DM855 in the expensive agglomerations for a family of four), a subsidy called Lastenzuschuß can be granted. The income of the household is still limited to the levels given above. This subsidy is also available for tenants, but is then called *Wohngeld*.

The monthly subsidy for each household receiving a grant averages DM150. In total, the subsidy costs the Bund and Länder nearly DM4 billion per annum, which is double the amount of subsidies for low-rent housing (Haus und Wohnung 1990: 56f; cf. Fig. 4.5 category "housing").

The Lastenzuschuß is given to 7% of favoured households, while 93% are tenants and so receive Wohngeld. Although the amount paid by the Bund and Länder to owner-occupiers is small compared with that paid to tenants, 124,000 households were subsidized in 1989 with the Lastenzuschuß, an amount that, in respect of social housing, was reached in the four years 1986-9 (Haus und Wohnung 1990: 47/58).

CHAPTER 5

The land-market process

5.1 Price-setting

The price paid for a plot of land depends on many factors, including the current economic and social situation, national and regional policy on urban development and the legal environment (planning), particularly at the level of the municipality. The location of a plot also determines its value (§194 BauGB, Dieterich 1990b: Rn 48).

The location of a plot, for example, its distance from the city centre, main roads and railways or green space, is of great significance. The relative importance of each factor depends on the intended use of the plot. In a residential area, for example, access to public transport or green space may be more important than access to main roads; easy access to the central business district (CBD) or main roads is more important for office location. Other factors include the size and form of a plot, the slope of the land, the condition of the soil (whether clean or contaminated) and the plot's relation to neighbouring buildings (Dieterich 1990b: Rn 52).

A plot's value is, however, most strongly influenced by the legal and planning circumstances. Figure 5.1 shows the correlation between land value and the stage of zoning, from farmland to totally developed building land. The stages are called *Entwicklungsstufen zum Bauland* (qualitative steps in the development process of a site). They are commonly used in property valuation and are currently given in the *Wertermittlungs-Verordnung* (WertV 88), a federal decree on land and property valuation (see, for example, Kleiber et al. 1991: 208ff). Because of this relationship, if a municipality decides to develop a site, it could generate enormous increases in land value to the benefit of landowners, but without their giving any service in return.

Although this pattern exists in Germany, there is no standardization of real market prices of land at different stages in the planning process. Nevertheless, a rule of thumb can sometimes be helpful (Rössler et al. 1990: 45): the *Bauerwartungsland* (land zoned in a preparatory land-use plan – FNP) is worth between 25% and 60% of the value of totally developed build-

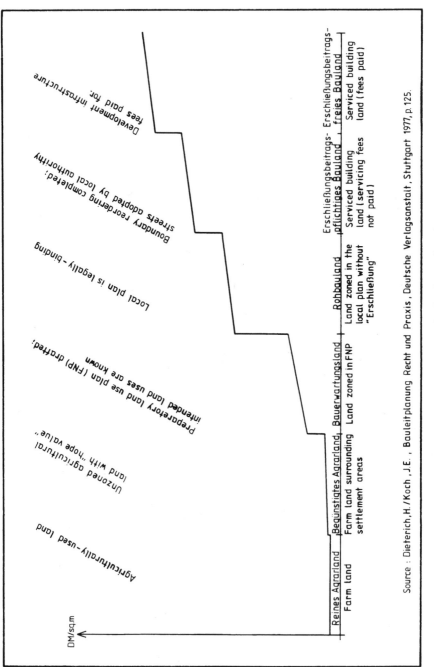

Figure 5.1 Relationship between land value and zoning stage.

Source : Dieterich, H./Koch, J.E., Bauleitplanung Recht und Praxis, Deutsche Verlagsanstalt, Stuttgart 1977, p. 125.

ing land (*Erschließungsbeitragsfreies Bauland*). The precise proportion depends on the timetable of development by the municipality and the degree of certainty that it will be carried out. The *Rohbauland* (land zoned in a local plan) is worth between 50% and 80% of the value of completely developed land. This percentage depends on the structure of ownership and boundaries, and whether procedures such as replotting (Umlegung) are necessary or not.

Other important factors include the *Art der baulichen Nutzung*, or land-use type (whether, for example, it is for residential or commercial use), which is fixed in a plan, and the *Maß der baulichen Nutzung*, or density of use described within the land-use plans. In investment-orientated developments, the latter factor is especially important next to location.

For building land in particular, the Gutachterausschüsse für Grundstückswerte (valuation committees) use real market prices to determine the Bodenrichtwerte (standardized land values). These land values are determined for all locations in a town, so that each square kilometre of developed land will have its standard land value. These values refer to the value of the land only, even in built-up areas. The Bodenrichtwerte are determined annually and are normally published in maps and in a report (§193 BauGB). They are derived from market values collected by the committees and are calculated for a standardized but typical plot in a common location. They are calculated for both residential and commercial areas, although they are more important in the residential areas and when public authorities are buying land. The values become more accurate the more purchase contracts they are based on.

Three methods are recommended by the Wertermittlungs-Verordnung for the valuation of undeveloped land and property. The comparative method (*Vergleichswertverfahren*) is considered the most accurate, using information about similar cases in the same location and giving values close to the market values. If possible, this method should always be used to determine the value of land in the context of the total value of an already-developed location.

The return method (*Ertragswertverfahren*) is applied in cases where the profit from property comes from rents. Typical examples are buildings with rented flats (normally more than two), shops and office-buildings.

The cost method (*Sachwertverfahren*) is used to value a building if the owner or investor needs it to produce something or to live in. Typical examples are owner-occupied single houses, factories producing special products, owner-occupied warehouses and public buildings.

With the increasing number of private investment projects, especially in the office sector (for example within business parks) a fourth type of valuation – residual valuation (*Residualverfahren*) – becomes useful (Kleiber et al. 1991: 716). Here, land value does not depend on the total value and the costs of the land, but much more on the profitability of the whole project. If the project is expected to be very profitable, the land will be more expensive than if the reverse is the case, assuming construction costs are stable and calculable. The most important factor is the ratio of floor to space.

5.2 Actors in the land market

Public authorities

The authorities of the Bund, the Länder and the municipalities act for the public on the land market, first as legal authorities and secondly as land-owners.

The legal issues and different instruments are described in Chapters 1.1 and 4.1, while Chapter 4.3 discusses the financial instruments that are important in the activities of public actors.

The municipalities are more significant as landowners than the Bund and the Länder. In this field the public authorities are dealing with civil law on such matters as contracts and hereditary leasehold. Land assembly by the municipalities is important in local policy-making and is often used in the development of industrial locations. Land assembly within residential developments is often connected to the implementation of social aims in the local policy (see the Stuttgart case study, Ch. 7.2). Public actors are less flexible than private actors in their land management, because of the regulations imposed by budget legislation. So in the recent period of increasing demand and prices, land assembly has become more difficult since other powerful actors, such as banks, can afford higher prices.

The case studies (Chs 7 and 11) contain examples of the behaviour of public actors: the influence of the Länder on regional planning (*Gebiets-entwicklungsplanung*) in Dortmund, the land assembly of municipalities in Hildesheim, Stuttgart-Stammheim and Dortmund, and the use of legal instruments in Frankfurt, Düsseldorf, Stuttgart and Dortmund.

The supply of land

The municipalities initiate development and create building land by zoning within their land-use plans. The owners of a greenfield development site are initially farmers or their descendants. It is rare for farmland to be owned by communities of joint heirs or even single heirs who are not engaged in agriculture. Other large landholders include the Church and the gentry.

A research project undertaken by the Federal Ministry for Spatial Policy, Housing and Urban Planning (BMBau) investigated types of landholders and changes in ownership during the development process of residential building land. It was modelled on five typical large towns and their neighbouring rural municipalities (Dieterich & Hucke 1985). The results must be carefully interpreted. Figure 5.2 shows the flow of sales during the three main steps

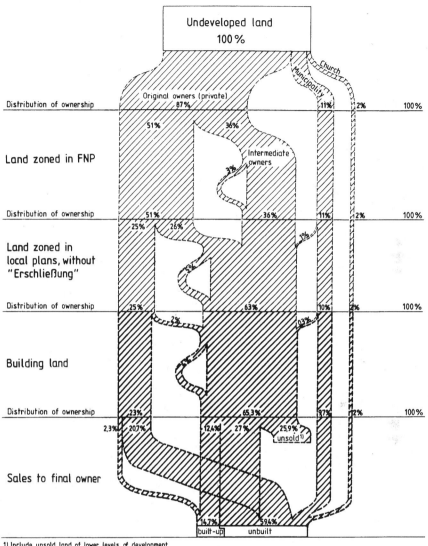

Undeveloped land
100 %

Distribution of ownership | Original owners (private) 87% | 11% | 2% | 100%

Land zoned in FNP
51% 36%
Intermediate owners
3%

Distribution of ownership | 51% | 36% | 11% | 2% | 100%

Land zoned in local plans, without "Erschließung"
25% 26%

Distribution of ownership | 25% | 63% | 10% | 2% | 100%
0,3%

Building land

Distribution of ownership | 23% | 65,3% | 37% | 2% | 100%
2,3% 20,7% 12,4% 27% 25,9% unsold[1]

Sales to final owner

14,7% 59,4%
built-up | unbuilt

1) Include unsold land of lower levels of development

Source: Scholland,R.: Die Bedeutung der Funktionsweise des Bodenmarktes.
In: Allgemeine Vermessungsnachrichten 1987, S.78.

Figure 5.2 Pattern of purchasers during the land development process.

of the process and the actors involved in the production of building land. Two-thirds of it was supplied from intermediate owners to the final owners, with less than a quarter coming directly from the original owners.

The municipalities and the Church kept their land (about 10% altogether)

109

in their ownership during the development process. They (as well as the original owners) normally offer undeveloped land to the final owners, often in the legal form of hereditary "long leasehold" (Erbbaurecht). This applies to the industrial land market, in which intermediate ownership by the municipality is much more frequent than in the residential sector.

Intermediate owners can be divided into three groups:

○ commercial intermediate owners, such as building contractors, market-orientated housing companies (*freie Wohnungsbauunternehmen*) or developers;

○ non-profit-making intermediate owners, such as housing associations, co-operative societies, public development companies (for example, the state-owned Landesentwicklungsgesellschaften, in Niedersachsen, Nordrhein-Westfalen or Baden–Württemberg, and the Deutsche Stadtentwicklungs-gesellschaft – DSK) and;

○ private intermediate owners (Scholland 1987: 77).

There is often more than one intermediate owner during the development of a plot of land. Intermediate owners who act commercially are most often involved when the development process is short, for example, two to three years. These owners normally build on the plot, and studies have shown that they have an 80% share of sales of developed plots to the final owners. During the 1980s, in comparison to the 1970s, it was more difficult to predict if and when a site would be developed, so the rôle of the commercial intermediate owners has decreased in importance. Consequently, their profits now stem mainly from the buildings themselves and less from the increase in land value during the development process (Scholland 1987: 80).

The non-profit-making intermediate owners act in close co-operation with the municipalities, who are often 100% shareholders in the companies. These intermediate owners are frequently directly involved in land policy. They mainly sell undeveloped land to the final owners.

Only 15% of all developed plots and 60% of undeveloped plots are purchased by the final owners from either their intermediate or original owners. Some 25% remain unsold for a longer time in one of the different levels of development, mainly owned by the non-profit intermediate owners (see Fig. 5.2, Scholland 1987: 80). The current trend is for a lower percentage of plots to remain unsold.

Speculation still occurs, and there are people and newly formed companies that invest money in farmland in the hope that it will be developed (see Ch. 6.7). This is a special group of actors working at high risk, often having extraordinary connections with local politicians. Today, private developments are more important than formerly, but developers do not normally embark on a development before it is certain. This means that the provisions for floor-space ratio and land-use in the Bebauungsplan must be predictable.

Land consumers

The consumers of building land are the first users or investors in the planned premises. The intermediate owners, as building contractors or developers, also demand land but not as final owners, so they are not considered to be consumers. Consumers must be distinguished for each market sector and there are further differences for owner-occupied and rented buildings. The patterns given below are not complete, as other combinations of actors are also possible, but these are the most typical. In general it is very important to identify the consumer for whom a development is being built (*Bauherr*) or being bought (*Erwerber*), as these are the only actors entitled to tax exemptions.

The housing sector (owner-occupation) In the owner-occupied housing sector it is common for a consumer to buy a building plot in a suitable location, intending to build a detached one- or two-family house. This is most frequent in rural regions (see the Hildesheim case study, project areas A and B, Ch. 7.1). The consumer is the Bauherr, who commissions an architect, asks for planning permission and commissions a building company to construct the house (Falk 1985: 105). If the project has a lower budget, then a prefabricated house or a "standard house" offered by building companies or architects may be chosen.

Another possibility is that the landowner, or Bauherr, will commission a building contractor who manages all the steps of the project. This involves, for example, the contractor asking for planning permission, subcontracting a building company and supervising the construction work (Falk 1985: 106). These contractors frequently offer "standard houses".

Another pattern is particularly common for the building of terraced houses and apartments. A building contractor or developer buys the land and looks for a buyer for the development. All details of the house, the transfer of ownership and the timetable are fixed in a contract. The Bauherr is the future owner. This method is becoming more common (see the case studies of Hildesheim, project area C, and Stuttgart-Stammheim, Ch. 7.1 and 7.2).

The housing sector (rented flats) In some cases, the Bauherr or landlord constructs a house with two or more flats, one of which he will occupy himself. The owner and landlord manages the renting of the other flats himself.

Private investment in the housing sector is common in the case of flats for rent and is promoted by tax incentives. High-earning people invest in the property market in order to reduce their tax liability (Ch. 4.3). These are the Bauherren and land consumers. The tenants are also important consumers, since the investment would not otherwise be profitable. Investors are often dentists, doctors, lawyers, architects, other self-employed people or small

111

companies. They normally invest in housing projects of between four and ten flats and sometimes they found companies with small numbers of shareholders. They normally authorize a company, such as an architects firm or a building contractor, to manage all aspects of the project. These companies are often later commissioned to manage the building (renting, maintenance, etc).

The demanders of building land for large housing blocks are the market-orientated housing companies (freie Wohnungsunternehmen), the non-profit housing associations (*gemeinnützige Wohnungsbaugesellschaften*), insurance companies and the large building companies. They behave in a similar way to building contractors in that they manage and supervise all aspects of the project. But they also initiate the projects and so are the Bauherren. At the same time they sell apartment buildings to private investors who then let the flats to tenants; sometimes the purchasers are even owner-occupiers (see the case studies of Hildesheim, Ch. 7.1, and Köln, Ch. 11.3).

This latter pattern is similar to the one described before, but here the risk for the investors or landlords is smaller because they are not the initiators of the project. The initiators act in a similar way to the developers. Housing associations and housing companies formerly often retained flats in their ownership; these companies own large building stocks from the 1960s and 1970s, and managing flats is an important part of their business today.

The industrial sector Owner-occupation is still the most usual pattern in the industrial sector. The consumers of land are the industrial employers who own and use the property (see the Dortmund case study, Ch. 7.3). In this owner-occupied industrial sector, problems exist both in managing extensions to existing premises and finding suitable new location. The local authorities provide their most important contact and support, with the office for the promotion of the local economic assistance board (*Wirtschaftsförderungsamt*) and the public estate office (*Liegenschaftsamt*). Estate agents are also often consulted. In the case of expansion, the employer normally solves the land problem by contacting the neighbouring landowners and the local authority. The construction of the premises is normally managed by the company itself, with a commissioned architect and building company.

In Germany, the investment market within the industrial sector is increasing and demand for rentable or leasable production space is rising. The projects mainly include more than 50% office space. They are managed by developers within a Gewerbepark-project (see the Düsseldorf case study).

The office sector Consumers of office space include large companies in the service sector that need thousands of square metres of office space for, for example, national or regional headquarters (for example, banks, and insurance, oil, energy, car and computer companies). These buildings are often owner-occupied, so the companies are consumers within the land market. However, private developments in this sector of the office market have recently increased (see the Frankfurt case study). Companies sometimes lease a building, or, if they only require a limited amount of office space, they are the main tenants.

Many offices are still in mixed buildings, with offices downstairs and flats or apartments on the upper floors. Smaller offices exist for companies with fewer employees. These projects are often managed by the same investors who offer flats for rent in the housing sector. Offices provide a substantial part of the owner's return.

The third group consists of investment projects for buildings including office space, the demand for which is rapidly increasing. Here the developers are the consumers of land, although the investors and tenants are no less important. Activities in this field are illustrated in the case study on Frankfurt (Ch. 11.2), while the case study on Düsseldorf discusses offices within a Gewerbepark (Ch. 11.1).

Property professionals

Various types of property professionals are involved in the land and property markets. They are not normally landowners. However, some of them, for example, developers or building contractors, are sometimes intermediate landowners, in which case they are counted as suppliers of land; this rôle has already been described.

Estate agents Estate agents act as mediators between other actors in the land and property market; they should avoid being intermediate property owners. They can be helpful in all areas of the land and property market, but are especially involved in two fields. First, they mediate between the original owners of undeveloped land and the (mainly private) actors who want to develop a site; this could be the final consumer or an intermediate owner (land-orientated activity). Secondly, they mediate between the Bauherr (owner of a building) and the user; this sometimes means looking for an investor, but is usually mediation between owner and tenants (property-orientated activity).

The legal foundation of the rôle of an estate agent is given in §34c *Gewerbeordnung* (the legislation governing trade and industry). The *Makler- und Bauträgerverordnung* (the federal decree on estate agents and building contractors) also applies. This states that the agent (or building contractor) needs permission from a public authority to be registered, must take out insurance and must keep accounts. These and some other regulations are designed to clarify and safeguard clients' interests. The criteria for granting permission are not onerous, in particular it is not necessary to demonstrate professional training. Agents are often businessmen or tax advisers.

The estate agent is only paid if his mediation is successful and results in a contract. Although there is legislation about negotiating the sale or letting of flats, the amount of commission is not fixed; it is normally 3–5% of the purchase price of a property (in both the residential sector and the commercial sector) or the equivalent of two–three months' rent. There is currently some debate about whether commission should be limited to two months' rent.

There are no exact statistics about estate agents in Germany. It is estimated, however, that there are about 11,000. Half of them, especially the large agent offices, are members of one of two unions:the *Ring Deutscher Makler* (RDM) ,700 members in 1985) and the *Verband Deutscher Makler* (VDM). For a long time, both have been interested in improving the legal foundations and the image of the profession. The RDM publishes monthly information about trends in the land and property market (*Allgemeine Immobilien Zeitung*).

It is estimated that estate agents are involved in between 30% and 40% of land and property transactions; these total about DM40 billion per year (Falk 1985: 112).

In general, most agents confine their activities to the local market. Here they are experts and are usually well informed about actors and market functions. Some large offices act in a regional or even nationwide context. Another group of agents specializes in one or two market sectors, the important ones being those in the industrial sector (Gewerbeparks) and in the office sector. They act on a regional or national level. This group increased in number with the growth of private development in Germany. In the above-mentioned fields they are involved in most of the transactions and offer an extensive service.

Developers Developers who manage private development projects are concentrated in the office sector and the high-tech industrial sector (see Frankfurt and Düsseldorf case studies, Ch. 11). Developers are still rare in the residential sector, although some building contractors or housing companies have sometimes undertaken developments. A developer is seen to be an intermediate owner of a plot who manages a project at his own risk, then sells the property to an investor either during the development or after its completion. However, sometimes the projects remain in the developer's ownership (see Düsseldorf case study, Ch. 11.1).

Many large German construction companies have expanded their businesses by offering special services necessary for property development. The construction phase within a project is carried out by the company itself.

114

Other developers do not employ their own construction workers and all tasks that they cannot undertake themselves are transferred to specialists such as architects and construction companies.

The developers act progressively. They canvass new office and Gewerbepark locations using their own initiative. The motivation comes, on the one hand, from the large amounts of money that will be invested and, on the other, from structural changes in a growing economy, in the service sector, for example. Within the property development process there is an obvious tendency to co-operate with the same group of associates. For example, a bank investing on behalf of its open-ended unit trust prefers to work with a special developer, who prefers to consult a limited group of estate agents and architects and a specific construction company (see the Frankfurt case study, Ch. 11.2).

Building contractors This group is very varied, as many building contractors are engaged in the housing sector but are also involved in other projects. They may belong to a group of developers. Contractors offer the whole service of managing and supervising a project commissioned by the Bauherr.

In an increasing number of cases, contractors arrange in a contract with a municipality to take over the public task of organizing the technical infrastructure (Erschließungsvertrag) (see Ch. 4.1). Often the same contractor manages the construction of the houses as a commission from the future owners, as described above. In principle this can be done by either small or large companies. There are also some public companies, mostly at the Länder level, who specialize in undertaking this task for small municipalities that have insufficient administrative capacity to do this for themselves, for example, the *Deutsche Stadtentwicklungsgesellschaft* (DSK) or the *Landesentwicklungsgesellschaften* (LEGs).

Building contractors in the housing sector normally act at the local level. Most companies are small, with only a few employees (such as in the Hildesheim and Stuttgart-Stammheim case studies, Ch. 7). Building contractors in other sectors, such as retail or public buildings, usually act at a regional or national level and are larger.

Construction companies Construction companies are similar to building contractors and they normally act at the local level of the property market. The number of construction workers per firm varies. Some are large companies with hundreds of workers and these act nationwide. The construction companies that operate locally normally employ 10–50 construction workers. In the winter the workers are often laid off for some months (see also Chs 1.4, 4.2, and the Hildesheim and Stuttgart-Stammheim case studies, Ch. 7).

Housing associations The rôle of housing associations (together with housing companies and private builders) is to guarantee a supply of flats. In the 1960s and 1970s they mainly offered flats for rent to low-income groups. In the late 1970s and 1980s they mainly offered apartments. An important part of their service is property management, because they own large building stocks. The former housing associations were non-profit organizations. This legally fixed characteristic was cancelled in 1990, so today these associations have to act under the same conditions as housing companies and private builders; in particular this means that they are now liable to pay tax.

There are about 1,800 housing associations in Germany. Of these, 1,200 are co-operative housing associations with 1.6 million members who build and maintain houses and flats for their members (Falk 1985: 85). They act at the local level of the housing market and are often limited to one or two city districts. Flats owned by co-operative associations are usually at the lower end of a town's price range. The other 600 housing associations are joint-stock companies with public shareholders such as the Bund, Länder, local authorities, trade unions or the Church. Most of the medium-size and larger towns have their own publicly owned housing association acting on the local market. The larger associations act at the national level.

The efficiency of housing associations in managing their building stock is often influenced by the current conditions of their staff and finances. Their flexibility with respect to enlargements in staff or capital is rather poor. On the other hand they have the advantage of close contact with local politicians and local authorities, as the Hildesheim and Stuttgart-Stammheim case studies illustrate (Ch. 7).

Two other important actors – investors and banks – are described in the context of the financial environment (Ch. 4.2). Instruments to implement plans are described in the context of the legal environment (Ch. 4.1) and the process is demonstrated in the three case studies of the operations of the land market (Hildesheim, Stuttgart-Stammheim and Dortmund) in Chapter 7.

CHAPTER 6

The outcome of
the urban land market

6.1 Demand for building land

The demand for residential building land should be seen in the context of the overall processes of the housing market, factors affecting the demand for housing, and changes in the structure of land ownership (see Ch. 1.4).

During the mid-1980s it was predicted that the demand for building land would decrease by the end of the decade and early 1990s (BMBau 1986: 60). This assumption was founded on expectations of a decrease in the population, although a continuing expansion of one- and two-family households, growing incomes and suburbanization had also to be taken into account. High vacancy rates for flats in 1985 and 1986 also supported this assumption.

The situation has now changed completely and demand in the whole of the land market is greater (GEWOS 1990b). This is mainly because of the rapid immigration of several hundred thousand ethnic Germans from eastern European nations and the former German Democratic Republic (see Ch. 1.3), which has led to shortages in existing housing stock and available building land. The problem was exacerbated by increased real net incomes after the mid-1980s. This new and unexpected demand is particularly for land that can be built on immediately. It is currently very difficult to estimate real demand since immigration levels are uncertain, as are changes in net incomes and the effects of relatively high interest rates (see Ch. 1.2).

Other factors have already influenced and will continue to influence demand for building land. A major factor is the increase in single- and two-person households and the clear wish of all types of households to enlarge their living space.

The demand for building land differs within regions. It is to be expected that in the countryside surrounding conurbations, because of better living conditions and lower land prices, demand for building land will be above average. Nowadays the southern German conurbations, such as München, Stuttgart and Frankfurt, are experiencing an increasing demand for land. Demand in rural areas is expected to remain below average.

Demand for commercial land is primarily influenced by the rate at which new companies are formed, and the movement of businesses within a municipality or between different towns. Over the past two decades, demand has been dominated by companies moving from inner urban locations to the outskirts or surrounding countryside to make use of adequate and fairly cheap building land for the expansion of production space. These trends still continue. In 1986, demand for land until 2000 was estimated at 22,000 ha or 1,470 ha per year (BMBau 1986: 68). Demand differs within regions and between different commercial uses. In the central city areas of conurbations demand is very high and is often dominated by international corporations. City-centre locations in particular are in demand from office-users, while production industries demand space in the surrounding countryside or rural areas. Locations with excellent traffic connections, attractive landscape design or a generally good environmental situation are favoured. In expanding cities the demand is often fulfilled by the provision of new business parks (see the Düsseldorf and Frankfurt case studies, Ch. 11).

6.2 The supply of building land

The supply of building land for residential use is characterized by the dilemma that, on the one hand, there is generally sufficient Bauerwartungsland (building land zoned by the Flächennutzungsplan), but that there is a growing shortage of *Bauland* (completely developed building land according to a local plan or an inner-area location – §34 BauGB) that can be built on at once. No current data are available.

In some locations, mainly the central city areas of large conurbations, the supply of Bauland has declined rapidly in recent years and demand cannot be completely satisfied. Nevertheless, reserves in the surrounding countryside and rural areas are usually regarded as sufficient (Güttler & Kleiber 1989). Only in the countryside surrounding prosperous growing cities such as München, Frankfurt and Stuttgart is the reserve of building land decreasing as land prices increase.

There are normally sufficient reserves of building land for commercial use, because of the active supply policy of the municipalities (see Chs 2 and 5). The reserves have fallen to a low level only in central city areas and expanding cities (see the Frankfurt and Düsseldorf case studies, Ch. 11). However, there is usually enough land in the countryside surrounding these areas. The situation is more difficult in old industrialized areas. There is normally enough derelict industrial land (for example, in the Ruhrgebiet conurbation the total amount of derelict land amounts to several thousand hectares of which only a small part has been decontaminated), but it is often not acceptable to large companies (see the Dortmund case study, Ch. 7.3).

The supply of building land is also determined by the amount of vacant or unbuilt single plots within an already-developed area. Sometimes these reserves total 35–45%, for example, in suburbanized zones of cities (Dieterich et al. 1991). In rural areas this share has remained high, while in larger cities it has decreased over the past few years.

It can be concluded that reserves are generally sufficient, although there are spatial variations. There is often the additional problem of building land reserves being unavailable or unusable.

6.3 Transactions on the land market

In 1988 there were about 70,700 contracts dealing with building land transactions at all stages of development, excluding agricultural land (Müller-Kleißler & Rach 1989). Within the approximately 600,000 purchases of all property types (GEWOS 1990b) the share of transactions of undeveloped building land came only to 12%. Table 6.1 illustrates the development of the number of purchases from 1980 to 1990, differentiated into the stage of development and type of use.

The largest category of building land, involved in 78% of all transactions and totalling 60% of building plot areas, is the *baureifes Land* (prepared land that can be built on at once). Purchases of this type of land decreased by an average of 3.5% per year in the early 1980s. Since 1985 the number of purchases has increased again, although it has not yet regained the levels of the early 1980s.

The decline of the second-most important category of building land, the Rohbauland (land zoned by Bebauungsplänen or binding local plans but not yet fully developed), is more obvious. The number of purchases decreased by approximately 11% each year until 1988, when it began to increase slightly.

Land for commercial use shows different trends. *Industrieland* (commercial land) is defined as unbuilt land which either belongs to companies and is to be used for expansion, or has a planned industrial use. After 1982 there was continual growth in both the number of purchases and the total amount of commercial land. This type of building land has become increasingly important and is expected to continue to be so.

The amount of building space and purchases differs according to spatial differences within and between southern and northern Germany. The decline was greatest in the less-populated countryside surrounding major urban centres and in rural areas. In these economically weak regions there was an annual decline in all purchases of about 8% and of 10% in area.

119

Table 6.1 Turnover of building land, 1980–1990, total number of purchases and as a percentage.

Type of building land		1980 %	1981 %	1982 %	1983 %	1984 %	1985 %	1986 %	1987 %	1988 %	1989 %	1990[1] %
completely developed	a	78.1	77.8	74.5	76.3	74.3	78.5	79.6	80.3	81.2	83.2	81.5
	b	62.6	62.7	62.6	61.9	60.0	54.3	56.2	54.5	55.3	59.0	55.6
zoned by local plan	a	14.2	13.8	12.2	11.3	10.5	9.2	8.5	8.1	8.0	7.1	8.2
	b	23.0	23.1	20.7	18.6	17.4	16.7	13.5	13.1	12.1	11.8	14.1
other building land	a	7.7	8.4	13.3	12.4	15.2	12.3	11.9	11.5	10.8	9.7	10.3
	b	14.4	14.2	16.7	19.4	22.6	29.1	30.3	32.4	32.6	29.2	30.3
land for commercial use	a	2.2	2.3	2.4	3.1	3.7	5.0	5.6	6.1	6.6	6.4	6.3
	b	10.4	10.7	10.1	12.5	13.3	19.0	22.1	23.7	26.8	25.4	24.1
land for traffic	a	5.0	5.5	10.2	8.8	10.9	6.5	5.6	4.7	3.4	2.7	3.1
	b	3.0	2.5	5.4	5.9	8.0	8.4	6.9	6.9	4.0	2.7	4.7
environmental space	a	0.5	0.7	0.6	0.5	0.6	0.8	0.7	0.8	0.9	0.6	0.9
	b	1.0	1.1	1.2	1.1	1.3	1.6	1.4	1.8	1.8	1.1	1.5
building land total (%)	a	100.0	100.0	100.0	100.0	100.0	100.0	100.0	100.0	100.0	100.0	100.0
	b	100.0	100.0	100.0	100.0	100.0	100.0	100.0	100.0	100.0	100.0	100.0
building land total (1000 ha)	a	97.6	79.2	78.8	73.2	66.2	62.4	62.2	62.6	70.7	81.9	71.0
	b	11.8	9.2	8.0	7.6	6.9	7.6	7.2	7.6	8.6	10.3	9.5

Source: Statistisches Bundesamt; Fachserie 17, Reihe 5

Note: 1/ provisional result.

2/ a = number of purchases.

3/ b = in space (ha).

Between 1980 and 1986, 56% of all transactions occurred in the southern states of Bayern and Baden-Württemberg and the state with the highest population, Nordrhein Westfalen. In 1986 at least 85% of all purchases of baureifes Land in Germany took place in small or medium-size towns of fewer than 50,000 inhabitants and the size of the transferred plots was above average. Villages with less than 2,000 inhabitants were particularly affected (Müller-Kleißler & Rach 1989).

In 1988 the average value of the turnover for all types of building land was DM7.5 billion, with an area-based price of DM88.25 per m². This compared with about DM3.1–3.6 thousand billion market value estimated by the Bundesamt für Statistik for all real properties including land, houses and flats, etc. (Euler 1991). Nearly 90% of the monetary value of the turnover for all categories of land came from Bauland.

Data for transactions with regard to the shares of occupiers and non-occupiers is currently not available.

The proportion of transactions made in the land market by different actors should also be taken into account. Private-sector transactions of building land in Germany have been consistently high in recent decades. Nowadays they amount to approximately 80% of all transactions on the land market. Buying

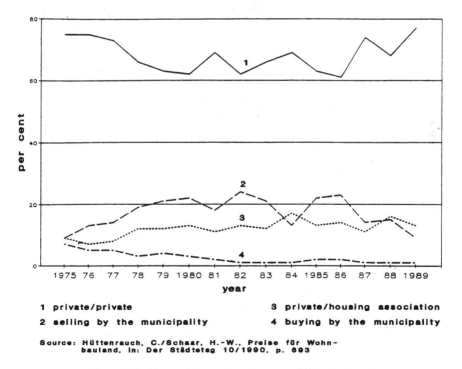

1 private/private

2 selling by the municipality

3 private/housing association

4 buying by the municipality

Source: Hüttenrauch, C./Schaar, H.-W., Preise für Wohnbauland. In: Der Städtetag 10/1990, p. 693

Figure 6.1 Share of transactions between different actors.

and selling with housing associations is less important and has not risen above 18% of transactions in the past 15 years. Nevertheless there are differences between cities. For example, the shares in 1989 in Dortmund were 28%, and in Stuttgart and München 40%, since housing associations in these cities have traditionally had a relatively high share of landownership.

In 1989 the municipalities' sale of their own plots was 9% of the total, while they are buying hardly any new land (only about 1% of all building land transactions in 1989). This shows that municipal land banking is becoming less important.

Although the turnover of unbuilt land is of great importance for urban development, its share in the property market as a whole is small, as mentioned previously. More transactions take place in built-up areas and the second-hand market is increasing in importance. In 1989, for example, the average share of transactions of unbuilt land based in approximately 20 large German cities was only 10% of all purchases. The share is normally greater if all types of cities and farmland are added.

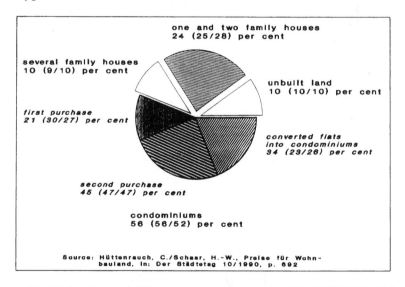

Figure 6.2 Market shares of different types of properties in 1989 (1988/1987) (20 largest cities).

6.4 Average sizes of building plots

The average size of building plots declined in the 1980s. Between 1980 and 1987 plots for single- or two-family houses decreased by approximately 100m^2 (2.1% per year) and by the end of 1987 the average size was 580m^2.

Plots for a group of family houses decreased by as much as 270 m^2 to an average of 960 m^2 (Müller-Kleißler & Rach 1989). In highly populated regions the relative decline was above average (see Ch. 7.2), while it was below average in rural areas, with plot size figures, for example, of 740 m^2 for single-family houses and 1,060 m^2 for multi-family houses.

The average size of plots for commercial use declined until the mid-1980s. Since 1985 it has stabilized or even increased slightly. In 1987 the average plot for production or storekeeping use was about 3,300 m^2, while plots for office use were about 3,150 m^2.

These trends appear to have continued, since land prices have risen and public authorities have supported the development of smaller plots in populated residential areas (see Ch. 2).

6.5 Prices

Following the explanation of land prices in Chapter 1, this section aims to provide more information about building-land prices in the different stages of the development process and within the spatial structure of regions.

The average price for baureifes Land per square metre increased by DM45 between 1980 (DM82) and 1990 (DM127), while Rohbauland growth was only DM18. Although there was a slight decline in 1985, prices continued to increase at a faster rate. The prices for commercial land showed similar trends. In prime locations, however, land prices have increased sharply in recent years.

The prices given in Table 6.2 should be used with caution, as the representative nationwide average prices are not very meaningful with regard to the large regional differences (refer to Figs 6.3 and 6.4).

German building land prices demonstrate a clear north/south divide in both the commercial and the residential sectors.

The price levels also differ according to city size. For example, in towns of above 500,000 inhabitants the price is on average eight times higher than in villages with fewer than 2,000 people. In towns of fewer than 20,000 people prices are generally below the national level, while in towns with a population greater than 20,000 prices are above average (Güttler & Kleiber 1989).

A different picture emerges if prices are divided into spatial regions and selected German cities. Between 1980 and 1987 the greatest increase in absolute prices took place in the central city areas of the large conurbations (DM108 per m^2) and in the densely populated surrounding countryside (DM64 per m^2). By contrast, prices in the economically weak rural regions were less than a quarter of those of the most expensive category. The increase in

Table 6.2 Land prices in Germany, 1980–1990, (DM per square metre).

Type of building land	1980	1981	1982	1983	1984	1985	1986	1987	1988	1989	1990[1]
completely developed	82.0	96.1	111.5	119.9	122.0	116.1	121.1	126.1	127.7	126.4	126.6
zoned by local plan	32.9	36.6	42.9	46.1	45.1	40.0	44.9	42.8	49.1	50.9	51.3
other building land	24.4	28.1	25.9	29.3	29.6	31.1	32.7	34.2	35.8	39.4	38.4
land for commercial use	27.0	30.5	32.2	39.0	39.0	40.1	39.5	41.1	40.4	42.3	44.9
land for traffic	15.1	17.0	15.3	11.0	13.5	12.3	12.4	12.2	12.1	14.6	12.2
environmental space	24.8	29.3	21.2	16.7	32.8	23.1	26.2	28.7	20.8	34.9	19.5
Total building land	62.4	72.7	83.0	88.5	87.7	78.7	84.0	85.4	88.3	92.1	89.2

Source: Statistische Bundesamt Fachserie 17, Reithe 5.

Note: 1) provisional result.

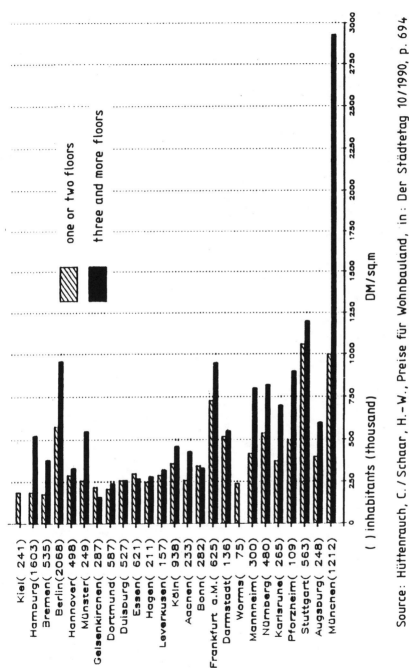

( ) inhabitants (thousand)

Source: Hüttenrauch, C. / Schaar, H.-W., Preise für Wohnbauland, in : Der Städtetag 10/1990, p. 694

Figure 6.3 Typical land prices for residential use in 1989 (DM per m²).

prices was greatest in the peripheral regions, with a rise of over 50% between 1980 and 1987. This indicates that suburbanization is taking place in these regions. It is expected that these trends will continue.

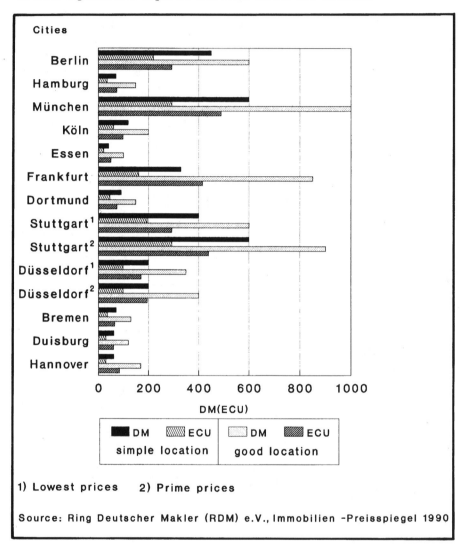

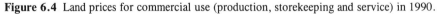

Figure 6.4 Land prices for commercial use (production, storekeeping and service) in 1990.

126

Table 6.3 Land prices of Baureifes land differentiated by spatial types of regions, 1980–1987 (DM per square metre).

Type of region	1980	1981	1982	1983	1984	1985	1986	1987
Regions with large urban agglomerations								
cores	236.5	293.4	313.4	326.8	369.0	346.4	343.2	344.1
highly populated surr-ounding countryside	143.6	170.8	198.2	199.1	207.2	205.0	202.2	207.8
less populated surr-ounding countryside	82.9	87.8	100.9	100.5	104.3	100.6	101.1	102.4
sub total,(average)	134.9	159.0	180.9	181.4	192.5	176.9	174.3	183.6
Regions with major urban centres								
cores	151.2	194.7	209.1	245.7	220.1	169.4	194.4	203.8
less populated surr-ounding countryside	56.0	66.7	77.7	82.9	83.6	78.4	81.3	84.5
sub total,(average)	63.4	75.7	85.8	94.0	90.7	85.4	89.6	94.3
Rural areas	53.0	64.2	76	86.1	82.8	74.5	77.0	86.4
Total (average FRG)	82.0	96.1	111.5	119.9	122.0	116.1	121.1	126.1

Source: Statistische Landesamter, Statistische Berichte M16.

6.6 Speculation in land and property

Information about the extent and distribution of speculation in land or property is not available. In Germany speculation is only registered with regard to the Spekulationsteuer (speculation tax, see Ch. 8). The information available from the Spekulationsteuer is unusable, as only purchases and sales within a period of two years are registered. Therefore only a few qualitative comments can be given.

The original landowners do not speculate on land. Also, hoarding land with the aim of inheriting properties cannot be interpreted as speculation, since owners who do this are not primarily interested gaining by speculation.

Speculation is used more often by some of the *Zwischenerwerber* (intermediate owners), although not all of them are speculators. Speculation on land normally takes place at all stages of the land development process (see Ch. 5.1 and Fig. 5.2). It is not uncommon to buy Bauerwartungsland (land zoned for building by a preparatory land-use plan) or farmland at the edge of the town and then to resell it after it has been zoned in a Bebauungsplan. The profit from the increased land values remains with the intermediate owners. Furthermore, because of continuously growing land values, the risk of speculating solely on baureifes Land has decreased and profits are greater, although land prices are usually high already. Nowadays speculation on land sometimes hinders the quick provision of land and does not help reduce the current building-land shortage.

Sometimes property professionals such as estate agents or developers urge public authorities to zone special land for their own interests (see, for example, the Frankfurt case study, Ch. 11.2). They normally promote their interests with investment projects they intend to implement on the plots. Nevertheless the profit from increased land values (or betterment) usually remains with the private companies.

Speculation on property usually takes place in two cases. In the past 10–15 years it has become more common to buy old houses (mainly in inner urban areas) and then to modernize and refurbish them. The houses or apartments are then sold to make a large profit. This process is mainly undertaken by professionals who often speculate with complete blocks or streets. This results in the conversion of rented flats into apartments and furthers the process of social segregation (see the Köln case study, Ch. 11.3). There are some statutory instruments to prevent such developments, but they have been unsuccessful (see Ch. 8).

The second type of speculation in real property is to buy buildings in districts undergoing rapid economic expansion (such as in the Frankfurt case study, Ch. 11.2). The "speculators" exploit the intense development pressure on these plots and gain exorbitant profits.

CHAPTER 7

Case studies of the land market

7.1 Hildesheim

This case study begins with basic information on the region and city of Hildesheim, which is used as the basis for a description of the conditions of the land market in the city, referring specifically to the development of the residential area of Marienburger Höhe/Itzum.

The geographic and economic background

Hildesheim is a medium-size municipality in northern Germany within the territory of the Land Niedersachsen. It is located 35 km southeast of Hannover, 50 km southwest of Braunschweig and 80 km west of the former east/west border. The centre of the region is Hannover, which is well connected to Berlin, Hamburg and the Ruhrgebiet. Hildesheim is within the EC region of Hannover (No. D32). There are many transport links to Hannover, to the world fair Expo 2000 and to Hildesheim itself.

Despite Hildesheim's proximity to Hannover, the city is not a member of the Verband Großraum Hannover. This is a regional planning association formed in 1962, by the city of Hannover and 20 neighbouring municipalities with more than 1 million inhabitants, during an expansive phase of regional development. For decades Hildesheim had been an intermediate centre. Recently the municipality has been classified as a higher-order centre by the Land government, despite its location quite near to the higher-order centre of Hannover.

In 1977 the formerly county-free municipality of Hildesheim was deprived of this status and was administratively incorporated within the county (Kreis) of Hildesheim (GA Hildesheim 1990: 6). Before German reunification Hildesheim was part of the West German *Zonenrandgebiet*, a politically designated area running along the former east/west border. Municipalities located within this zone, which was economically disadvantaged were able to claim additional federal funding (see Ch. 4.3).

In 1990, Hildesheim had a population of 104,800. This had increased in

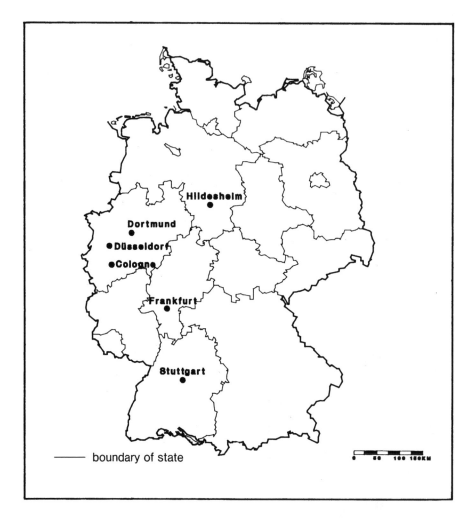

Map 2 Federal Republic of Germany: location of case studies.

the mid-1960s as a result of national economic trends. In 1974, nine smaller municipalities were incorporated. After 1974 the population decreased slightly and then fluctuated around 100,000. Between 1987 and 1990 the population increased by 4%. Currently, 6.5% of the population are immigrants.

In 1990 the city district of Itzum, which contains part of the residential area described in the case study, had a population of about 6,000. Located in the southeast of Hildesheim, it has been growing since the last decade, particularly between 1983 and 1990. In 1990 the population continued to grow and in January 1991 reached 6,458 (Stadt Hildesheim 1991a).

The municipality of Hildesheim had a built-up area of 24%, with 40% in agricultural use, 20% woodland, 10% taken up by the transport network and 6% used for social amenities (Stadt Hildeheim 1991a: 20). The industrial areas are concentrated in the north and east of the town. The preferred residential areas are in the south.

Within the past few decades the region of Hannover and the city of Hildesheim have been less affected by the economic recession than other northern parts of Niedersachsen. Development in Hildesheim is generated in the shadow of Hannover. However, it has been structured differently from that economic regions in southern Germany.

Between 1979 and 1987 employment in the service sector increased by about 35%, while employment in primary industries and construction decreased by about 35%. In 1989, 51% of all workers in Hildesheim were employed in the tertiary sector, 44% in the industrial sector and about 4.4% in agriculture and forestry. The main employers in the industrial sector are in processing and capital goods (Stadt Hildesheim 1989).

About 29,100 persons commute into the city, so that almost every second person working in Hildesheim does not live there. By comparison, only 8,100 persons commute from Hildesheim to Hannover, despite Hannover being the most important centre of employment.

In Hildesheim the number of workers in the agricultural sector is close to the national average, but this sector is especially important for the land market in the north of Hildesheim. It is one of the most productive agricultural regions in Germany with one of the best-quality soil types.

The land and property market
Land and property prices Land and property prices in the Hannover region follow the south/north differentiation, with higher prices being general in southern Germany. For the past 10 years Hannover has been one of the group of larger cities that offers relatively low land and property prices. In 1989 prices for building land in the region (excluding Hildesheim) fluctuated around DM280 per m^2, which is comparable to prices in places such as Bremen, but lower than in Hamburg (DM370 per m^2) and Münster (DM340

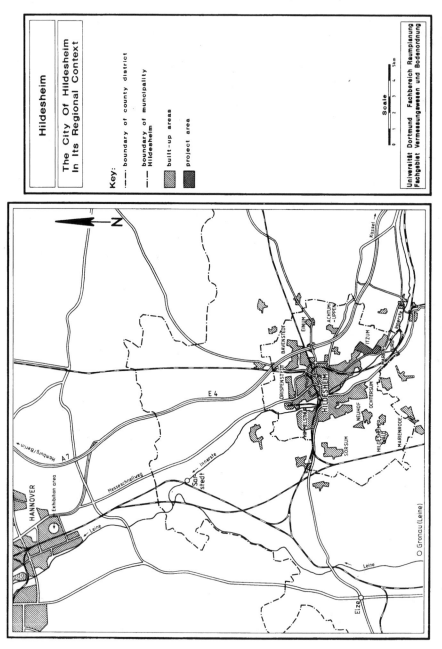

Map 3 The city of Hildesheim in its regional context.

Universität Dortmund Fachbereich Raumplanung
Fachgebiet Vermessungswesen und Bodenordnung

Hildesheim

The City Of Hildesheim
In Its Regional Context

Key:

—·—·— boundary of county district

—··—··— boundary of muncipality
Hildesheim

▨ built-up areas

▨ project area

Scale

0   1   2   3   4   5km

per m²) (Hüttenrauch 1990: 694). However, price differences between the city of Hannover and its surrounding suburbs are not as great as in southern German conurbations.

The price level in Hildesheim is lower than in Hannover, but the pattern of development has been similar during the past decade (GA Hildesheim 1990: 5). Prices paid for building land in 1989 reached a maximum of DM310 per m², which was paid in the most preferred residential district in the northern part of Galgenberg. On average, prices for residential building land are about DM150 per m², excluding development fees (GA Hildesheim 1990: 33). Prices of industrial building land are less than half those for housing land (see Fig. 7.1).

Prices in the south of the city are generally higher than those in the north. Over the past four years land prices increased by a moderate 5–8%, but they are currently rising rapidly.

In Hildesheim current prices for apartments are about DM2,300 per m² of living space, which is similar to Hannover. However, there has been a decrease of about 30% since 1983, when prices were DM3,300 per m². Since 1991 prices have increased again.

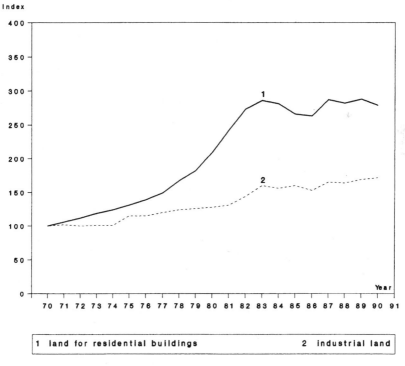

Figure 7.1 Development of land prices in Hildesheim.

The pattern of activity in the Hildesheim land and property markets is typical for a rural district. By comparison with the conurbation of Hannover, purchase activity in Hildesheim is low (see Table 7.1). In 1990, an average of 3 in 100 inhabitants of the municipality of Hildesheim participated in the property market (GA Hildesheim 1990: 8).

Table 7.1 Turnover on the land market in Hildesheim (number of purchases).

	1985	1986	1987	1988	1989	1990
building land	630	605	674	677	738	977
built-up land	680	801	864	879	949	947
condominiums	314	280	322	326	380	429
turnover total	2,215	2,269	2,292	2,524	2,715	2,985

Source: Municipality records

Building land supply and demand Until the end of the 1970s there was substantial allocation of land for multi-storey housing in Hildeseim. By comparison, during the mid-1980s a significant number of such dwellings were unoccupied and free to let in Hildesheim, as in many other parts of Germany. In the mid-1970s, there was great demand for building land for single-family houses, and this demand was met by the supply of large building plots. In the 1980s, the case-study area of Marienburger Höhe/Itzum contained almost the entire supply of Hildesheim's residential building land for single-family houses, as well as for houses for rent.

There has recently been insufficient provision of privately owned building land open to immediate development without implementing specific long-term planning instruments. Currently 30 ha of privately owned building land has been allocated in the southern city district of Ochtersum. Development will begin here, on the western side of the river Innerste, in 1993. Furthermore, the extension of the Marienburger Höhe/Itzum area to the east (35 ha) is already included in the Flächennutzungsplan (FNP) and will be developed in the future (Stadt Hildesheim 1979).

Demand for building land in Hildesheim is comparable to Germany as a whole. During the 1970s it focused on rented properties and, because of the lack of dwellings, housing development in Hildesheim was dominated by multi-tenant projects and often managed by local housing associations. Demand for single-family houses increased after the 1970s as personal incomes rose. During the 1980s, demand for rented flats decreased further because of the economic and structural problems of northern Germany. Until 1990 there was little construction activity in this sector. In recent years, however, the demand for multi-tenant houses has increased slightly.

Demand for individual housing projects increased in the 1970s, and so building activities mainly involved single-family houses and apartment developments. By the end of the 1970s, demand for 500 such building plots had been received by the government of Hildesheim. Increased demand for this type of housing led to the development of the Marienburger Höhe/Itzum area at the beginning of the 1980s.

The city's land policy Even though Hildesheim has a high-order status, this does not mean that it is entirely independent of the neighbouring conurbation of Hannover. Accordingly, housing and land policies in Hildesheim are affected by economic and political decisions taken in the context of the development of the Hannover region. For example, the intensity of housing activities over the past few decades, and particularly the increased activity following reunification, demonstrates this relationship.

In the past the agriculturally dominated land-use structure, combined with a relatively flat countryside, favoured the potential allocation of large settlement areas at locations within the city boundary. The local authority traditionally preferred to allocate one large development area that would provide sufficient supply for several years. This is how the allocation of land in the case-study area of Marienburger Höhe/Itzum occurred.

The local authority enforced a land-assembly policy designed to give it such large areas for disposal. This was easy to implement when fulfilling the demand for industrial building, but was more difficult to implement for residential building land. Over the past decade, banks and insurance companies have purchased more land for capital investment, for example, in the current and future development areas west and east of the case-study area.

In the past, non-profit-making housing associations provided living space for households on lower incomes. However, at present no local policy exists to facilitate the acquisition of property for households on lower incomes by providing special subsidies.

The project areas of Marienburger Höhe/Itzum The housing development area of Marienburger Höhe/Itzum is located 5 km southeast of the city centre and east of the Marienburger Straße, which links Hildesheim to the municipalities to the southeast (see Maps 3 and 4). The site links the built-up areas around the university (former college of education) with the city district of Itzum, and half of the area belongs to this city district. The project area totals 114 ha and is more than 70% of the settlement reserves allocated in the current preparatory land-use plan of Hildesheim. The area is well situated, facing south on a slight slope. Today it is one of the favoured residential areas of Hildesheim.

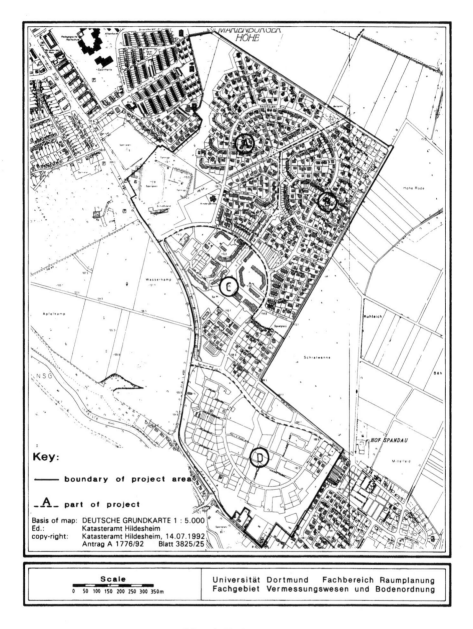

Map 4 Project area.

The area is served by three main access roads that connect twice with the Marienburger Straße and once with the neighbouring city district of Itzum (see Map 4). Besides the circular access roads, separate traffic-restricted roads and green areas with paths for pedestrians provide interior access. The streets are constructed in different ways and with different widths, depending on their importance. Access roads have a total width of 13.4–17 m and include space for pedestrians, cycle routes and lines of trees on both sides. The traffic-restricted routes have a total width of 4.5–6.5m and are paved, not "tarmacked". Buses go to the city centre four times an hour.

The northern part of the development area is dominated by large single-family detached houses built on large building plots of equal-size. In the south, the main building types are terraced houses and multi-storey houses. Some shops are concentrated in the middle of the area. The proportions of the whole project area (114 ha) developed for different land-uses are (Stadt Hildesheim 1991b):

Building plots for housing	60.5%
Streets	16.0%
Public space (green areas, playgrounds, social amenities, etc.)	17.0%
Sports area	6.5%

The relatively large figure of 17% for public green space and playgrounds includes the obligatory noise barrier along the Marienburger Straße and a protection area 42m wide (see Map 4). No housing is allowed here, and so the area is used as a public amenity and it forms part of the footpath and cycle network. The noise barrier running along the Marienburger Straße (4m5 high, 25 m wide and 800 m long) is required by environmental law to reduce the noise levels inside the site.

Intensive development of the area occurred at the start of the 1980s when demand for single-family houses increased. This is also why public and private building contractors were engaged in owner-occupied housing and apartments. The northern parts of the area (A and B) were the first to be developed, followed by the southern parts (C and D). The development has almost been completed and construction work will continue until 1993 in district D only. The development is legally based on four local plans and on the implementation of replotting of land (see Map 4 and Fig. 7.2).

The process of development

In 1974 the northern part of the Marienburger Höhe/Itzum area (60% of the total area) had been bought by the city of Hildesheim, under the authority of the Land Niedersachsen, for the expansion of the college. The land was paid for by the Land government. Although the college extension finally was not carried out, the land was sold to the local authority at a low price.

Over the past decade the city has sold the building land located in the northern part mainly to private individuals, who have used it to construct free-standing single-family houses. Building plots owned by the city in the central area were sold in the main to private contractors, who had to build

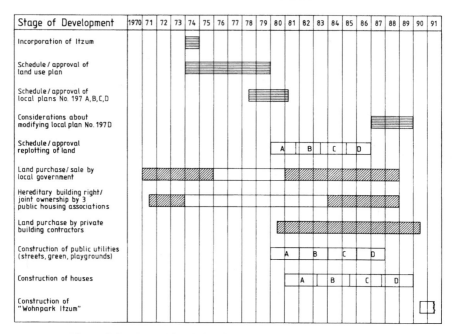

Figure 7.2 Time-table Hildesheim – Marienburger Höhe/Itzum.

terraced housing and multi-storey houses on them. Some plots, including plots owned by the city in the southern part, were sold to insurance companies.

In the 1970s, three local housing associations, the Gemeinnützige Baugesellschaft zu Hildesheim (GBG), Gemeinnützige Wohnungsbaugesellschaft für den Landkreis Hildesheim (KWG) and Beamten-Wohnungsverein zu Hildesheim, acquired hereditary leasehold for the other half of the southern part (see Map 5). The associations preferred joint ownership in order to reduce the risk of an unprofitable investment. After the replotting of the land was completed in 1986, each of the three associations owned a similar building site and the hereditary leasehold was partly continued and partly transferred into individual ownership (see Map 6). Before 1974 the then self-governing municipality of Itzum intended to develop land in the northwest of its territory and so approved a local plan. After the incorporation of Itzum, the intended development did not go ahead

The housing associations sold the building land to other investors in accordance with their financial and personal situation and with demand. Finally, in 1989, they sold large shares to private individuals, as well as to a private contractor who is currently constructing a complex of multi-storey buildings named Wohnpark Itzum (see below). One half of the southern part was privately owned farming land. The private owners, most of them farmers, either sold their property or used it themselves.

In July 1980, because of increased demand, the city administered a specific valuation system for the allocation of city-owned building plots for sale for the development of individually built owner-occupied houses. At that time, the number of applicants (915 in 1980) was four times greater than the existing number of building plots. The system took account of different criteria, including marital status, time of application, personal handicap and current residential situation. During the distribution process of the plots in the Marienburger Höhe/Itzum area, more applicants were satisfied. There were fewer restrictions on the last building plots because the number of applications had rapidly decreased (Stadt Hildesheim 1980).

Legal instruments

After 1974, when the worldwide oil crisis brought about national economic recession, plans for further expansion of the college were scrapped. The city re-allocated the area for residential use. The Marienburger Höhe/Itzum area was designated for residential development because of the ownership structure, the south-facing hillside and good transport connections.

Because the area was large, the local authority divided it into four parts covered by four current local plans, named Nr 197A, B, C and D. These give a structure to the area by providing three different types of housing (Stadt Hildesheim 1981). The boundaries of the four local plans are equivalent to the parts A, B, C and D already described and shown on Map 4.

Nr 197A has 17.07 ha of building land and provides land for detached housing with one or two floors, on building plots of 550–650 m². It is a traffic-restricted residential area. The local plan was approved in January 1981 and it covers an area to the northwest of the project area. Houses with one floor have to be constructed as free-standing buildings, and houses with two floors may be built as terraced houses. The proportion of houses was determined at GRZ 0.3 (acreage area ratio) and GFZ 0.5 (floorspace ratio). Traffic connections are provided by an access road running through the area and small routes passable for pedestrians and cars. Parking lots have to be integrated into private building plots. Account has to be taken of a high-tension electricity cable passing through the plan area.

Nr 197B, has 15.1 ha of building land and provides for the same house type as district A. The same regulations for the proportion of houses and traffic requirements apply. The plan was approved in December 1980 and the area is located in the north, near to district A.

Nr 197C provides three- and four-storey houses inside the Hansaring, close to some shops and offices. Here the proportion of buildings was determined at GRZ 0.4 and GFZ 0.8–1.1. The plan area is situated to the south of A and B and was approved in July 1980. This has been targeted as the infrastructural centre of the whole development area. The other portions of the plan area will contain types of houses similar to those in A and B. In addition row houses with a floor-space ratio of 0.8 were allowed. A compulsory noise barrier along the state road L491 is included.

Nr 197D, with 21.2ha of building land targets multi-storey houses and some free-standing houses. The proportion of buildings is set at GRZ 0.4 and GFZ 0.5-1.1. In 1987, influenced by public housing associations, the city planned to limit the acreage and floor-space ratio so that single-family houses could be constructed. However, at the time there was discussion about providing land for rentable housing in addition to the existing allocation for free-standing owner-occupied houses. The discussions took account of the changed situation on the housing market, especially after the opening of the former inner-German border. As a result, the proposal to allocate land to single-family houses instead of multi-tenant buildings was cancelled in December 1989. The allocation of multi-storey housing remained. The privately owned Wohnpark Itzum has been under construction since last year and it provides condominiums in four-storey buildings. In addition, the local plan caters for houses with one or two floors without restricting the length of the house complexes. The four-storey buildings currently under construction (Wohnpark Itzum) are obstructing the view of the countryside nearby. Owners of single-family houses located at the boundary of the local plan Nr 197D, which already have been constructed directly in front of the Wohnpark Itzum, are considering claiming damages from the local government. The decision to erect their houses on this site was dependent on statements from the local government to modify the neighbouring local plan Nr 197D from multi-storey housing to free-standing housing.

No modifications to local plans Nr 197A, B and C have been necessary, which significantly shortened the time needed for their approval. In September 1981, the only regulation approved was a design regulation concerning type and colour of roofs, ensuring that architecture be similar to that of existing neighbouring buildings. In July 1987, an equivalent regulation was approved for local plan area Nr 197D.

The local plans only lay down a few restrictions on building types, and several different types of building, construction materials and designs could be used, according to the individual ideas of the home-owners. In sectors A and B in particular, no homogenous architecture has been developed. The division of the area into four local plans was not really necessary, since the plans came into force at the same time. It would had been possible to combine all settings and restrictions into one comprehensive local plan. The design standard applied to the streets and green areas was not laid down in the local plans, but was imposed by the local government and implemented during construction work.

The project was implemented using the land replotting process. In Hildesheim, a specific type of replotting named *Wertumlegung* is normally utilized. The area was subdivided into four parts, almost equal to the territory of the local plans Nr 197A, B, C and D. The process can be best explained using the local plan D. Map 5 shows the quantity and size of the plots at the beginning of 1980. Map 6 illustrates the result of the replotting procedure and the distribution of ownership and building plots after implementation in 1986.

The administrative decision for replotting the land in the Marienburger Höhe/Itzum area was taken in 1980 and the preliminary stage of the replotting procedure was completed in 1986. As a result of this procedure the local plan area Nr 197D was subdivided into large preliminary building-land blocks. This was possible because each of the landowners already owned large sites before the replotting started. Each of these blocks contained enough space for 10–12 single building plots. Following the replotting procedure, these blocks were subdivided into single building plots according to the plans of the landowners. This had been achieved by the end of 1990.

Each landowner had to make 33% of his undeveloped land available for public amenities within the area, including streets (16% of the whole area) and green space (17%). The remained area was divided into favourably shaped building plots and distributed among the landowners. Each new plot had to have access to public streets and was categorized as developed building land with a higher land value. The local authority was entitled to levy a charge on the landowners involved, based on the resulting surplus value of the redistributed building plots (see Ch. 4.1).

The formerly undeveloped land bought during the replotting procedure was given a uniform value of DM30 per m^2. The land values of the new building plots varied depending on the GFZ stated in the local plan:

GFZ 0.5: DM46 per m^2
GFZ 0.8: DM50 per m^2
GFZ 1.0: DM54 per m^2
GFZ 1.1: DM57 per m^2

The city of Hildesheim did not aim to use the replotting of land to its own financial advantage. Accordingly, the surplus value is calculated with reference to the costs resulting from the procedure, except the provision of land required for streets and green space. These procedural costs (especially surveying costs) were DM1.80 per m^2. However the surplus value was a minimum of DM16 per m^2 and a maximum of DM27 per m^2, so the municipality did not exercise its legal right to charge the full betterment levy (surplus value). This underlines the local authority's policy of not taking maximum financial advantage of replotting. The city was nevertheless able to conduct the procedure without any financial difficulties because the land

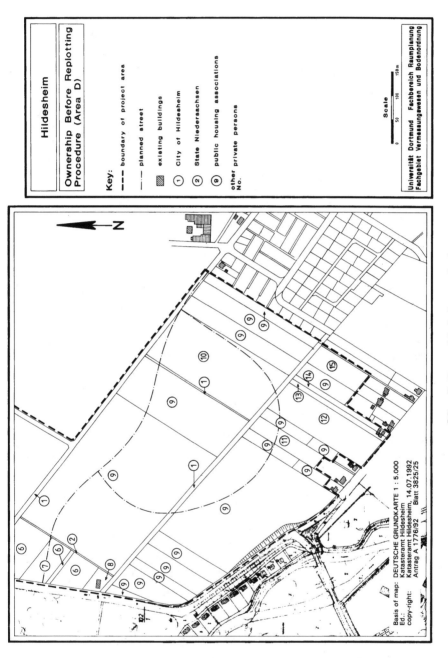

Map 5 Ownership before replotting.

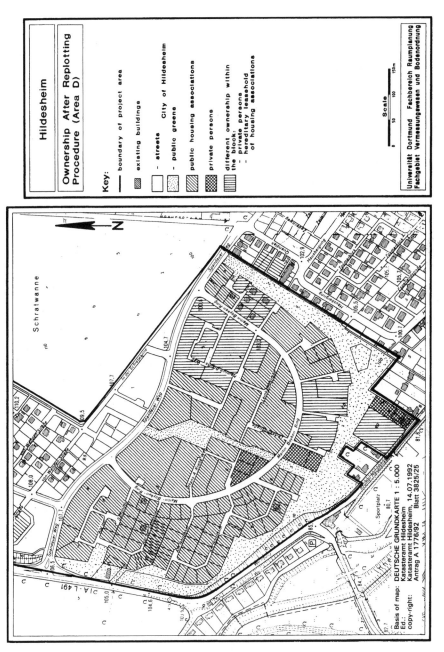

Map 6 Ownership after replotting.

involved that was in the ownership of the city had been bought earlier at very low prices.

Discharge contracts (*Ablöseverträge*) were used to charge development fees for the areas covered by the four local plans. Landowners pay fixed development fees after they have purchased their building plots and before public utilities are fully installed. The first payments of the development fees for building plots were due in July 1981. The construction of services started in the northern parts of the area in October 1980 and was finished in 1987.

The total amount of the discharge includes the cost of different factors in the development. The cost of each factor was determined by the city. Apart from the usual development implemented by replotting, the cost of the land needed for streets and green space is not included in the cost of public utilities (Erschließungsbeiträge), because the land was taken without compensation within the replotting procedure.

○ Costs for streets, ways and public squares within the territory of local plans Nr 197A, B and C were estimated at DM18.351 million. The development fee charged came to between DM34.5 and DM50.0 per m², depending on valid acreage ratio (GRZ) and floorspace ratio (GFZ).

○ Costs for streets, ways and public squares within local plan area Nr 197D were estimated at DM15.88 million. Landowners had to pay between DM41.00 and DM64.50 per m².

○ Development costs for access roads in the whole area were estimated at DM3.987 million, the amount to be paid by landowners was DM5.50–8.00 per m².

○ Development costs for green spaces, noise protection and playgrounds, which had to be financed by the landowners, were estimated at DM4.995 million, or DM7.00–10.00 per m².

○ Costs for the sewerage system also differed, varying between DM9.15 and DM15.45 per m².

In summary, development fees for each landowner ranged between DM56.00 and DM97.00 per m² depending on the GRZ and GFZ allowed for each building plot. The municipality covered 10% of the costs (§129 BauGB). The charges for developing and servicing the plots are rather high, especially considering that the costs of the ground are not included. However, the public utilities are large-scale and of good quality and design.

The actors and their behaviour

Besides the activities of the local government, other actors were involved in the development of the area. Three public and several private building contractors, some insurance companies and private home owners were involved.

Public housing associations During a period of 16 years (1971–87) public contractors owned building land in the southern part of the project area but no building activities occurred on this site. As explained above, implementation took place from the north (district A) towards the southern borderline of the project area (district D).

The completion of the replotting procedure led to building plots being made available for the housing associations, which were fixed by local plan at a GFZ of 0.8. But only few plots in their ownership were developed by the public housing associations themselves. For example, the public housing contractor *Gemeinnützige Baugesellschaft Hildesheim* (GBG) owned about 8.1 ha land within the project area. A large proportion of this land was transferred to the city (1.4 ha) and to private persons (4.2 ha), leaving only 2.5 ha to be developed by GBG itself.

The situation was similar for the other two housing associations. They sold parts of their property as building plots, built up part as free-standing single-family houses on behalf of future home-owners, and on other parts sold single-family houses constructed by the associations themselves as terrace houses.

This behaviour was typical of housing associations in the 1980s, although they had been mainly engaged in low-rent housing in earlier years. Because of the decreasing rental housing sector in the 1980s and the government policy of favouring owner-occupied single-family houses, the housing associations were compelled to vary their behaviour and to act in a way similar to private contractors. The public contractors sold parts of their original property depending on their own market investigation and willingness to take risks. Also their financial and personal capacity was an important factor determining the extent of their engagement. It must be noted that the capital of the public housing associations is concentrated significantly in many older, cost-intensive, low-rent blocks of flats. Managing these flats is their main task, new development is an additional activity. Thus, money for additional investments is less readily available. Also, because of staff limitations, these publicly owned non-profit organizations do not operate in a flexible manner. They prefer to keep a plot for a long period of time than take a heavy risk. Although the demand for rental flats has increased in recent years, the housing associations in Hildesheim are still continuing to develop only single-family houses. However, there is also a strong demand for these premises. The housing associations did not make a loss within the area, although their house prices are slightly higher because of the financing costs for the land over 16 years.

Private building contractors Three private contractors, two regionally orientated and one locally orientated, carry out building activities in the area.

The private contractors were mainly involved in the southern part of the area, where they purchased plots from the city and from private persons. Thus, both the regional contractors had intermediate ownership on some plots. Their involvement depended on current demand, which kept their investment risk low. One of them constructed condominiums intended as capital investment for private persons. The local contractor offered single-family houses and owner-occupied terrace houses. Single-family houses were constructed in agreement with the later buyers and the row houses were constructed according to the standardized product of the contractor.

The private contractors employ subcontractors, and do not employ their own construction workers. Normally, only a small management staff is employed. Even when the number of orders declines, there are still operations going on that provide employment for management staff but not for construction workers.

In addition, large insurance companies that operate on a national level were engaged in the Marienburger Höhe/Itzum area, but only to a limited extent. They constructed a few multi-storey projects in district C and sold them as condominiums. These companies normally manage only multi-storey projects. They even did this for a time in the mid-1980s, when housing associations hardly engaged in this type of building.

One of the regional private contractors constructed the Wohnpark Itzum. This development will consist of four-storey buildings, arranged as four single complexes within a green area. Residents will enjoy an unrestricted view of the surrounding countryside.

Each complex consists of five segments with eight condominiums in each. The living space in each dwelling will be between 48 m^2 and 98 m^2. It is planned to construct each complex within one year, and the first two complexes are currently under construction. The building costs for the Wohnpark overall are estimated to be DM40 million.

After final completion the Wohnpark Itzum will provide 220 condominiums which are to be built in four phases:

1 complex (44 condominiums) September 1991
2 complex (60 condominiums) March 1992
3 complex (60 condominiums) May 1993
4 complex (60 condominiums) May 1993

The prices for condominiums provided in the first complex were calculated on the basis of DM2.750 per m^2 of living space. In 1991, after completion of the second complex, prices will rise to DM2.850 per m^2, which is relatively high for Hildesheim. Higher prices are under consideration for the fourth complex of condominiums. Most of the condominiums will be rented out at DM11–12 per m^2 of living space. In comparison to other multi-tenant houses in Hildesheim these rents are very high.

146

The Wohnpark Itzum is an investment project. The private contractor took the risk of engaging in multi-storey housing development on the assumption that high-income households would generate increasing demand for condominiums in such buildings. As mentioned above, no building activities of this type had occurred in the city for years. The developer wants to sell the condominiums in blocks of eight flats to private investors. The assumption was proved valid before construction was complete. All 44 condominiums in the first complex had already been sold by December 1989. Three investors bought one housing segment each, the contractor himself kept ownership of one, and 20 condominiums were sold to different investors. Furthermore, all 60 condominiums in the second complex were subject to preliminary purchase contracts by May 1991, although at that time construction had only just started. Most investors are from Hildesheim and 80% of them do not intend to live in their properties. Investors are motivated by tax advantages (see Ch. 4.3). A special information brochure and an expert opinion on tax advantages had been offered by the contractor in promoting the Wohnpark, giving potential investors detailed information. Management of the dwellings and insurance services are offered as a package by the development company, providing extra financial return on the project.

In 1989 the private contractor purchased the 20 ha location at a price of DM125.00 per m^2 from one of the public housing associations. The local authority gave administrative support to the implementation of the project. Purchase contracts were drawn up in December 1989, before the proposed modification of the local plan was again cancelled. The city government assured the private contractor of the reallocation in favour of multi-storey housing, and the necessary building permissions were granted by the authority very quickly in July 1990, allowing construction on site to start in August 1990. After 16 years without building activities, the development of the Wohnpark was accelerated by all those involved.

The city of Hildesheim created an additional financial aid programme, as well as the tax advantages, in order to increase rental housing provision within the city. This occurred during a period of strongly increasing interest rates in early 1990, because both the politicians and the developer were afraid that the project would otherwise fail. This financial programme focuses on the 220 condominiums of the Wohnpark Itzum. Buyers of condominiums may apply for a grant of DM4 per m^2 of living space monthly provided by the city over three years, whatever the status of the tenant and level of rent. That means, for example, that the owner of a 70 m^2 condominium will receive about DM10,000. The only condition of the programme is that applicants have to rent out their condominiums to third parties.

This kind of direct subsidy for landlords is not common, and it demonstrates the commitment of the municipality to reinforce rental housing.

Because there are no social policy conditions concerning rent levels or tenants, these subsidies may be regarded as a kind of gift to the investors. The subsidy, however, does seem to have been an important factor in ensuring successful implementation of the first part of the project.

Construction companies In Hildesheim there are a few large construction companies employing as many as 100 workers. The large building projects undertaken in the past in Hildesheim (for example the restoration of the old marketplace and the erection of the new local savings bank) were implemented by large construction companies. Many of the smaller companies went bankrupt during the 1970s, when orders were scarce, and the remaining small private companies are engaged only in owner-occupied family houses.

However, the development of the residential area Marienburger Höhe/ Itzum offered significant employment opportunities, especially for the smaller companies. They were commissioned by housing contractors and by private persons. The latter normally only require the services of small companies for the construction of their own single-family houses.

Three medium-size companies were the main contractors in the Wohnpark Itzum. Normally only a few units are constructed by one single company, but nevertheless in the Marienburger Höhe/Itzum project much of the terrace housing in the middle part of the area was built by single contractors. This allowed construction works and materials to be standardized, although options for individual interior modifications were available. Depending on additional modifications, prices were in the range DM228,000–300,000, excluding plot costs.

Private owners Most of the single-family houses covering large portions in the north of the Marienburger Höhe/Itzum area were financed by private persons with moderate to high incomes. For example, for free-standing single-family houses prices of between DM500,000 and DM800,000 were often paid. If the building plots were purchased from the city, the private owners were required to develop them within five years of purchase.

Contractors were not involved, architects and construction companies having been contracted directly by the home-owners. Open-market conditions were modified by favourable tax treatment available for private persons by investing in housing projects. The larger houses typically stand on large plots, and normally they will be expected to stay in the ownership of the occupier for a long period.

Most of the owner-occupied row houses in the middle part of the area have been purchased from contractors, as mentioned before. Apart from favourable tax treatment, public funding relating to income levels (or social housing) did not often apply. Nevertheless, because of financing problems and migration, the rate of change of occupier is much higher than in single-family houses. However, most of the terrace houses are owner-occupied.

Condominiums provided by the Wohnpark Itzum are attractive as capital investment for private individuals and to reduce their tax burdens. It is likely that most of them have no housing problems.

7.2 Stuttgart

This case study describes the typical market conditions and processes that occur in the housing land market in the city of Stuttgart, which is one of the largest conurbations in Germany. The land market in Stuttgart is characterized by high prices for all types of land-use. Background information about Stuttgart is given first, followed by details of a typical example of housing development, the residential area of Stuttgart-Stammheim-Süd.

The regional context

Stuttgart is the ninth largest city in western Germany and is the major city within the regional conurbation of Mittlerer Neckar. It is the capital city of the Land Baden-Württemberg and is the Land's economic, administrative and cultural centre. It belongs to the EC region Stuttgart (Nr D 81). The city covers an area of $207 \, km^2$: a somewhat small area because there have been no incorporations since 1942, unlike most other large cities in the former Federal Republic of Germany. The city is divided into 23 city districts. The northernmost city district of Stammheim is one of the main development areas for housing in recent decades and is the location of the residential area described in this case study.

Stuttgart is connected to the motorway system by two main feeders, the federal roads B10 and B27, which ensure good accessibility to the city (Map 7).

The topography, characterized by rolling hills separated by small valleys, is an important feature limiting settlement development. The basin-shaped valley of Stuttgart is open only to the north. In addition to the specific topography, significant local emission concentrations, mainly caused by traffic congestion, frequently lead to smog concentrations. Because of restricted air circulation, serious discussions are currently going on about the wisdom of building additional office blocks in the city. Despite this, it is regularly said that the residential and recreational value of Stuttgart is high (Hintzsche 1990: 267). This positive image came about because of the agricultural landscape surrounding the city.

Stuttgart is surrounded by 26 towns, some of more than 70,000 inhabitants. This conurbation forms one of the most powerful economic regions of

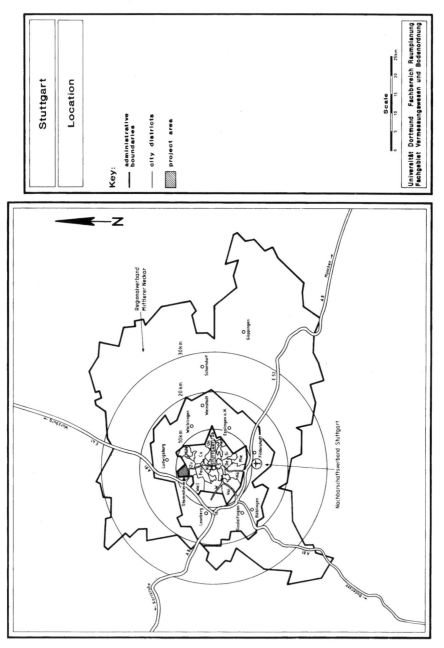

Map 7 Stuttgart: districts and boundaries.

Germany. Stuttgart and the neighbouring towns have established an inter-communal planning association called the Nachbarschaftsverband Stuttgart (Map 7). This institution is empowered to conduct planning procedures for land-use plans and landscape plans for each municipality. Its members are also members of the Regionalverband Mittlerer Neckar, which is a larger regional association and includes 179 municipalities. Since 1973 the Region-alverband has been responsible for the preparation of comprehensive regional plans, but not for their financing and implementation (Dierks & Kirchner 1989: 13). The term region as used below is equivalent to the territory of the Regionalverband.

The geographic and economic background

Population and settlement During the past few decades, the favourable employment market in the Stuttgart region has attracted inward migration from Baden-Württemberg as well as from other parts of Germany. The population of the region increased continuously to reach a maximum of about 2.36 million. Since then population within the region has fluctuated around 2.3 million people. This is in line with other conurbations in southern Germany and indicates a positive net north—south migration (BMBau 1990a: 61).

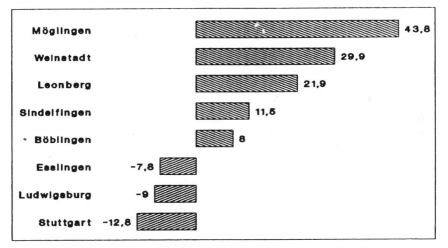

Figure 7.3 Gains and losses of population in the Stuttgart region 1970–87 (%).

Another demographic pattern within the region is that of a population shift to the city periphery, and beyond it to small and medium-size towns. Between 1970 and 1987 suburbanization led to a population loss of 13% within the city, while the suburban area gained 10% (Fig. 7.3). During the same period, regional population overall inclined 3.8% (IHK 1990: 13). In small and medium-size towns located between the growth axes as determined

151

in the regional plan population increased by about 24 % (Hecking 1988: 63).

Despite population stagnation since 1973, settlement development increased continuously during the same period. The highest intensity of development shifted from the city to the suburbs. From 1968 to 1985 the settlement area in Stuttgart increased by about 27% (equivalent to 21 km^2) and within the other parts of the region by about 47.5%, which is equivalent to an additional 121.7 km^2 (Hecking 1988: 53).

The development of the settlement structure in Stuttgart follows different patterns in the north and south of the city. In the 1950s and 1960s the first industrial areas were established in the northern districts, such as Feuerbach and Zuffenhausen. As a consequence, residential areas were dominated by multi-tenant buildings of the type required by blue-collar workers, which were provided by the city. By contrast, major portions of the southern territory were developed after the 1960s, with a more varied mix of buildings. City districts such as Vaihingen and Möhringen in the south are still the most preferred residential areas in Stuttgart, and very high land prices are found there.

In 1989 the population of the city of Stuttgart was about 559,700, having reached a maximum of 633,200 in 1973. This decline was caused by the process of suburbanization, and the loss of population continued until 1985. Since then, population has increased slightly, but between 1973 and 1989 a decline of at least 11.6% occurred.

Stuttgart's current population density of 2,700 per km^2 is one of the highest in Germany, but in the district of Stammheim, the location of the case study, it is currently 4,764 persons per km^2. Stammheim has always had many inhabitants because of its proximity to the industrial areas. Continuous housing development increased its population by about 8% between 1970 and 1989 from 10,390 to 11,242 (Stadt Stuttgart 1990a: 8).

Economy and employment The regional economic base is currently one of the strongest in Germany. Capital goods industries such as electrical and mechanical engineering, and various growth industries, ensure good results in both quantitative and qualitative economic output. As a result of profitable markets for products and high productivity, the region's gross product of DM39,300 per person in 1986 is equivalent to that of regions such as München (DM44,360 per person) and Frankfurt (DM46,420 per person) (IHK 1990: 45). The region of Stuttgart houses the headquarters of well known enterprises such as Daimler-Benz, IBM-Germany, Porsche and Bosch, which all indicate the region's economic strength.

Total employment in the region increased between 1970 and 1987 by about 16%. The loss of employment in the industrial sector was about 11%, while employment in the service sector increased by about 58% (IHK 1990: 19).

In 1989, total employment reached 375,200. Of these, 108,500 were employed in service-related industries. In 1989, the industrial sector overall accounted for 37.9% of all employment (140,800 employees). The tertiary sector comprised 62.1% of employment (230,400 employees), which was concentrated in the south of Stuttgart.

The regional unemployment rate of 4.4% in 1988 was still the lowest in Germany, for example, the München region recorded 6.2% and the Ruhr region 15.1% (IHK 1990: 17).

Besides administrative, social and cultural facilities, the majority of the workplaces are provided in the centre of the region, in the city of Stuttgart. In 1987 about 210,000 employees were commuters, while only 30,000 people commuted out of the city. Most of the incoming commuters (83%) live in rural districts around the city, the most important of which are shown in Table 7.2 (Greiner 1991: 17).

Table 7.2 Commuters to the city of Stuttgart.

County	Commuters to Stuttgart
Ludwigsburg	57,200
Esslingen	48,600
Rems-Murr-Kreis	40,200
Böblingen	27,300
Göppingen	5,600

Source: Municipality records.

The land and property market

Supply and demand of available land The current preparatory land-use plan, the FNP, which was approved in 1974, assumed a supply of land of 115 ha for residential use and 203 ha for business uses. Not all plots that should have been allocated became available, mainly because of the lack of legally binding local plans and constraints hindering the planning procedure for the local plans.

There is no exact current data on available land reserves (Stadt Stuttgart 1974). In 1988 a study based on the preparatory land-use plan approved in 1974 estimated reserves of land zoned for housing at 178 ha but only 27% of this land is available immediately (Güttler 1989: 239). The sites ready for development without planning hindrance are already occupied, and supply is therefore scarce. Reserve land for business uses has been estimated at 90 ha, 56% of which is available immediately and 44% of which will be available within the next five years (Güttler 1989: 241). Assuming a yearly demand of about 10 ha, this should be sufficient for about 10 years.

In Stuttgart, the supply of building land has always been restricted:

○ In the Stuttgart region the limits for settlement development compatible with environmental pressures have been achieved. This is especially true within the central city, where about 47% is built up.

○ The allocation of residential building land in the common master plan of the Nachbarschaftsverband gives a high priority to smaller municipalities, because allocation will not generate any deviations for the plan. In addition, the city of Stuttgart traditionally provides only small shares of immediately available building land in order to decrease resulting costs for the provision of services.

○ The people in the Stuttgart region are reluctant to sell land.

Nevertheless demand for building land in Stuttgart is very intensive and the high land prices described below are the result. Besides high income levels, which result in increased demand for living space per person, there are additional factors causing the demand in Stuttgart:

○ Workplaces, infrastructure and cultural events are centralized in the city. Neighbouring municipalities provide much less infrastructure. Many well qualified white-collar workers with high incomes demand attractive residential locations nearby or within the city.

○ People value the ownership of property at the highest level; ownership of land is seen as the most important status symbol for individual social standing.

○ Realteilung is very common. This is the process whereby, in the case of inheritance by several heirs, one piece of land is divided into individual plots, which leads to sites that are too small for development. As a consequence, the original inhabitants have a small piece of land and make attempts to expand their ownership.

Turnover on the market In recent years there has been a great increase in land prices and turnover in different market sectors. In general the Stuttgart land market is characterized by a high turnover of building land and housing property, with nearly 6,000 sales in 1990. This means that every year 0.8% of the city area comes up for sale. The intensity of market activity mainly results from Stuttgart's rôle within a much larger regional land market (GA Stuttgart 1990: 10).

In the Stuttgart land and housing market, apartments are the main sector. In recent years turnover of building land and built-up land have fluctuated at steady levels (Fig. 7.4). Turnover of multi-tenanted buildings decreased, because former opportunities for converting these into apartments were restricted by a revised law.

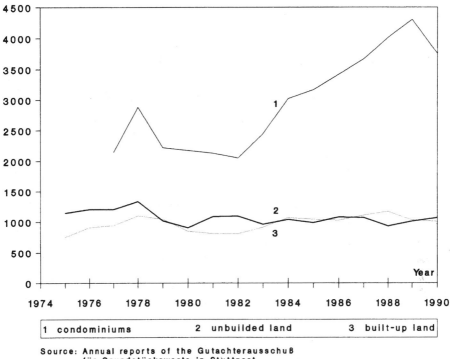

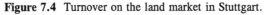

Source: Annual reports of the GutachterausschuB
für Grundstückswerte in Stuttgart

Figure 7.4 Turnover on the land market in Stuttgart.

Land and property prices In 1990 the price for residential building land in Stuttgart reached between DM1,400 and DM1,500 per m², twice as much as the level 10 years previously. The level of prices established in recent years, as shown in Figure 7.5, is one of the highest for large cities. In 1989, for example, prices in München were DM1,212 per m² and in Frankfurt DM625 per m² (Hüttenrauch & Schaar 1990: 691). In 1990 prices for industrial land in Stuttgart rose to DM1,000 per m² in certain cases and, on average, to DM750. Traditional locations of the manufacturing sector within the inner city are being increasingly replaced by businesses in the service sector that are able to pay such high rents (*Stuttgarter Zeitung* 2 May 1991: 25).

Prices for apartments have increased continuously in the past two decades. Between 1978 and 1990 they doubled from DM2,550 to DM4,905 per m² of living space. The only phase of stagnation occurred between 1984 and 1986, which is in line with the decrease in turnover described above, while in most parts of Germany prices fell (GA 1990: 22). With reference to the different housing market sectors (Fig. 7.6), price development is characterized by increasing levels with the exception of a few short periods of stagnation (Hintzsche 1990: 266).

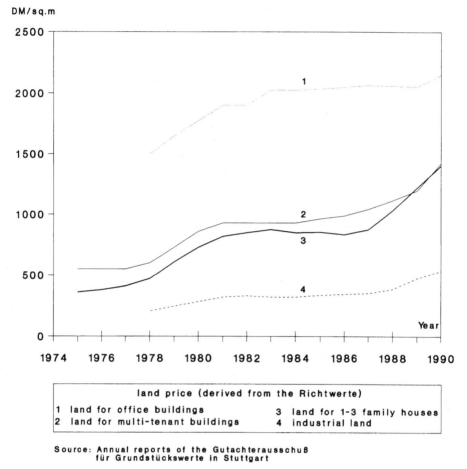

Figure 7.5 Development of land prices in Stuttgart (in absolute terms).

Rents for dwellings built since 1979 are about DM10.2 per m² and the average level for rents in Stuttgart is between DM8 and DM10. Rents are estimated excluding all service charges (water, heating and gas) which will raise actual occupation costs for tenants (GA 1990). In addition, the first renting of new dwellings is very expensive: DM20 per m² is not uncommon.

Land prices in municipalities located around Stuttgart are also increasing. In 1990 very high land prices were paid in Esslingen (DM1,200 per m²), Ludwigsburg (DM1,000), Böblingen (DM1,200), Filderstadt and Sindelfingen (DM1,050) (Stadt Stuttgart 1991: 1). Even in less-attractive city districts such as Stammheim, land prices for residential building land of about DM1,000 per m² are typical (Fig. 7.7).

Besides residential attractiveness, new public passenger transport lines significantly influence price levels in the municipalities around Stuttgart. Travel-

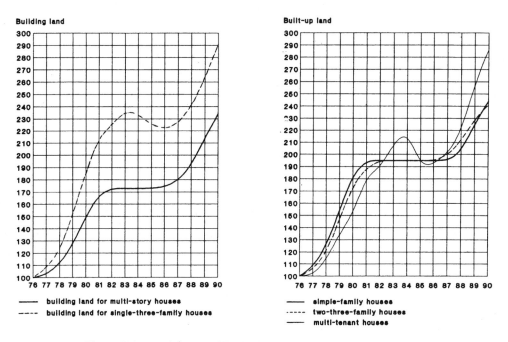

Figure 7.6 Development of land prices in Stuttgart (in relative terms).

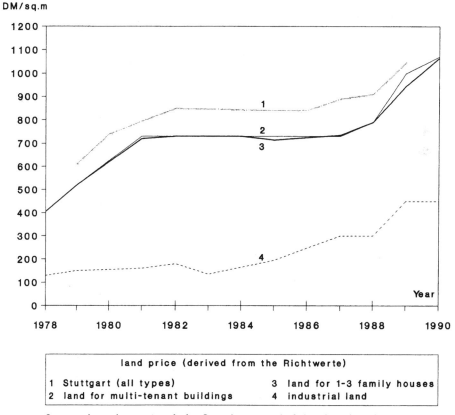

Figure 7.7 Development of land prices in Stuttgart–Stammheim.

ling times of up to one hour into the city are accepted by commuters, so municipalities directly connected to Stuttgart by public transport have become preferred residential areas, with subsequent rapid increases in land prices (Stadt Stuttgart 1991).

By comparison with other conurbations in Germany, there is no obvious explanation for the extremely high level of land prices in Stuttgart. Certainly economic attractiveness that has been sustained over recent decades is one of the important factors. But, for instance, economic development in Hamburg is also favourable, without producing land prices that are so high. Certainly the topographical situation, which restricts outward expansion, must influence price levels. But this limitation is not restricted to Stuttgart.

The housing sector Since the beginning of the 1970s, social housing has decreased in the inner-city districts in line with national trends. Between 1968 and 1987 about 10,000 new dwellings subsidized by public funds were constructed. In 1968, 16.7% of the building stock (35,002 dwellings) was publicly funded; in 1987 the share was only 14% (37,242 dwellings). This is because of a decrease in new public housing projects and the substantial losses resulting from public housing dwellings being privatized. This change of housing status is legal after a specified period of time. Nearly 40% of all social-housing dwellings were built between 1949 and 1958, so the social-housing stock will continue to decrease (Stadt Stuttgart 1990a).

During the same period, the number of apartments increased significantly. While in 1968 only 5.6% of all dwellings (11,638) were apartments, in 1987 this had reached 15.4% (41,553) (Hecking 1988: 61). This illustrates the influence of the high land prices that predominate in Stuttgart. The construction of apartments is more profitable, because the expensive building land needed for an apartment is much smaller than that needed for the erection of single-family houses. Public-housing development was thus reduced because significant subsidies from local government would have been required to cover high land prices.

The dominant type of building in the region varies. Between 1980 and 1985 in the suburbs about 90% of all building permits per year were for single-family houses, because of the process of suburbanization. In Stuttgart, only 55% of such building permits issued each year were of this type (Hecking 1988: 61). After 1979, many of the newly constructed residential buildings were erected in the district of Stammheim (Table 7.3). In this district the number of dwellings increased steadily, although development of the largest site was delayed until 1978. Today, 15% of all existing dwellings are social housing (Stadt Stuttgart 1990a: 137).

Land and property policy The lack of available land because of high land

158

prices in Stuttgart has meant that subsidizing owner-occupied housing projects is an important part of municipal policy, in addition to the state government grants available. Only 27% of households were owner-occupiers in 1985, so the municipality promotes this type of ownership through the establishment of programmes designed to enable families with moderate incomes to become owner-occupiers. Under open-market conditions, such families would not be able to finance the acquisition of building land. This was the main reason for establishing the development plan for the project area in the district of Stammheim as described below. With the exception of specific housing projects, there is no comprehensive policy to decrease land-price levels.

Table 7.3 Dwellings in Stuttgart Stammheim.

Year	Number of dwellings
1956	1,870
1961	2,536
1968	3,002
1987	4,632
1989	5,007

Source: Stadt Stuttgart 1990a:137.

Because of high land prices, the local government of Stuttgart is unable to buy and allocate enough building land to satisfy current demand. Accordingly the municipality does not own enough land to influence land prices. A stringent land-banking policy would be too expensive, but in order to extend development areas, voluntary replotting of land is very often used (von der Heide 1989). This depends on the widespread process of Realteilung, described above, which is very often responsible for undevelopable and awkwardly shaped pieces of land within the development areas. In Stammheim, the development of the site depended heavily on the replotting of land.

The number of replotting procedures in the city of Stuttgart has decreased since 1970. In 1970, 30 official and 48 voluntary procedures (*Freiwillige Umlegung*) were operating with respect to the redistribution of residential areas. At this time, local government tried to force voluntary replotting, which was advantageous to the city, as government staff were less involved. Over the years, approximately 335 ha were privately replotted by voluntary action (Stadt Stuttgart 1990b). Because of 1983 land-transfer tax legislation that withdrew exemption from land-transfer tax in the case of voluntary replotting, the number of voluntary activities dropped rapidly (10 in 1977, 5 in 1983, 4 in 1987 and none in 1989). So local government concentrated again on official procedures.

The local authority transfers of city-owned land are seen as an additional source of funds for the city's budget. A large proportion of developed city-owned sites has been transferred to the SWSG (Stuttgarter Wohnungs- und Siedlungsgesellschaft), a local housing association and a wholly city-owned institution. The city housing office manages a database for cases of urgency, where families with housing problems are registered. In 1989, 4,500 people were registered.

Description of the project area of Stammheim-Süd

According to the current preparatory land-use plan (Flächennutzungsplan) of Stuttgart, Stammheim-Süd was one of the last large residential areas that could be developed. The 27 ha site is located within the city district of Stammheim, to the north of Stuttgart. The city centre is approximately 8 km away (Map 7).

The area was developed as a traffic-restricted residential area in which housing development took place on relatively small building plots and at a high density. Building activities on this site followed on from the development of a larger housing area located at the northern and eastern boundaries. These housing areas were built two decades ago. On the west of the area there is a smaller industrial area.

The area is currently characterized by free-standing houses and some multi-storey houses in the northern part, and by terraced houses erected by public subsidy on former city-owned sites in the south. Despite being public housing, the latter are of a moderate building standard in well landscaped surroundings. The area is especially attractive for families.

Main traffic connections are through the B10, a federal road, and by a public passenger transport line. The B10, running on the south side, is one of the main transport links between the city and the suburbs and also a main feeder to the motorway system. There is no direct access by the B10, which is separated from the site by a noise- and pollution-protection wall. Vehicular access to the interior of the area is provided by the Korntaler Straße. Secondary roads run perpendicular from the access road into the interior of the site. Besides these there are paths designated for pedestrians and cyclists providing access to home entrances. Common parking lots are concentrated both in sections of the secondary roads and as underground garages in the noise-protection wall. The area is easily reached by trams and buses. Open space and some small playgrounds are concentrated along the southern boundary.

Development of the area would not have been legally possible without the construction of the noise-protection wall. In order to decrease traffic density in the neighbouring residential areas, the B10 and also the B27, another federal road passing the site, were extended between 1970 and 1980 under an official approval of a plan implemented in 1969. After the extension was completed in 1980, traffic on the B10 was expected to be about 45,000 vehicles a day (Stadt Stuttgart 1980). The resulting noise levels within the project area had to meet given standards required by the pollution-environmental law (Bundesimmissionsschutzgesetz). In order to decrease noise levels, noise protection was the only possibility. It is constructed in the form of a natural wall with pedestrian areas on top. On average the wall is 10 m high and 50 m wide. It was constructed at the same time as other building activity in the area.

The development of the residential area of Stuttgart-Stammheim-Süd occurred over two decades. Two main factors delayed the development of the area until 1978 (Fig. 7.8). First, a weakness in the city's budget at this time hindered approval of the local plan until 1978. In the 1970s there was strong housing demand in Stuttgart and the city started to allocate building areas in different locations, depleting the city's budget. The financing of installation of public utilities, which must be partly paid for by the city, was therefore not assured. Secondly, federal environmental law was enacted in 1974 that lead to uncertainty over local government jurisdiction with reference to the construction and financing of the noise-protection wall.

The development process

Ownership By the 1960s, the municipality of Stuttgart had begun to acquire large portions of land in Stammheim. Prices for land ready for development were DM55–240 per m^2. By this time the development of this former agricultural land for housing had already been targeted and a draft local plan had been determined. Because of Realteilung, many plots could not be developed because they were too small in size or irregular in shape (Map 8).

The implementation of the project was based on using the Umlegung instrument (replotting of land), which resulted in developable, favourably shaped building plots for each landowner (Map 9). The large portion of city-owned sites in the project area (62%) in 1970 was not a typical pattern for housing development in Stuttgart. The city gained great advantages from its landownership. Under the Umlegung procedure, following its completion by 1983, the distribution of property among the landowners was almost the same as at the start of the procedure. Streets and public green space came into public ownership. After completion of the Umlegung, most of the plots owned by the city were transferred into hereditary leasehold. In following years, mainly after 1986, the city started changing hereditary leasehold into ownership. The city sold built-up land to private home-owners at favourable prices. This does not reflect market value under open-market conditions. Market values of building plots within the project area were influenced by transactions between private persons, the highest turnover occurring in 1984. The maximum level achieved for building land was DM675 per m^2 in 1984.

161

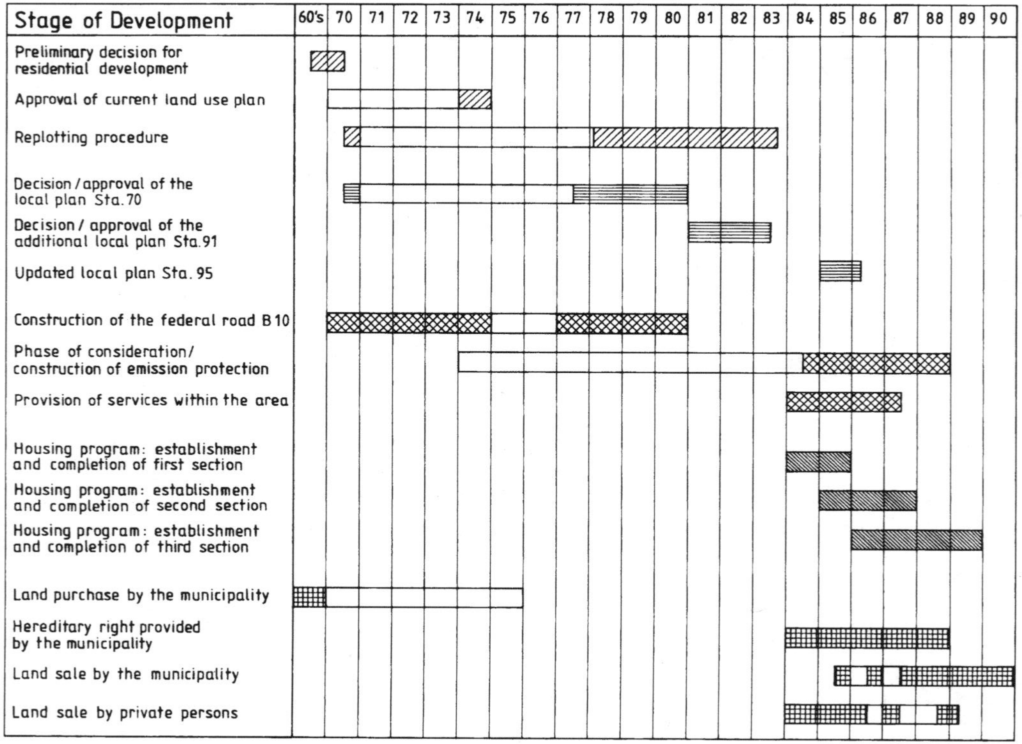

Figure 7.8 Timetable: Stuttgart Stammheim-Süd.

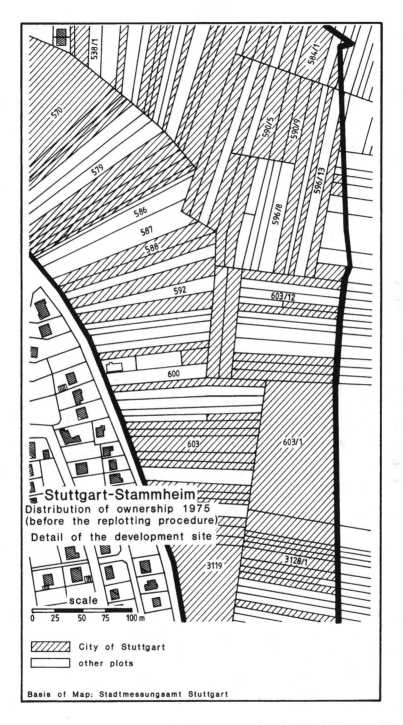

Map 8 Stuttgart-Stammheim: detail of development site, showing 1975 ownership.

Map 9 Stuttgart-Stammheim: distribution of ownership 1983.

The local plan   The current local plan for this area originated from a decision about the applicability of a local plan in July 1970, and a decision for implementing the replotting of land procedure was set as a result. After 1978 the draft plan of 1970 was revised in the three following stages:

1. The first plan (Sta 70) was approved in November 1980, with the exceptions of the later noise-protection areas. Construction of single-storey and multi-storey residential houses was possible. The northern part was targeted for free-standing buildings and the southern portion was to be developed with terraced houses on small building plots. The proportion of buildings was determined at GRZ 0.25 (acreage area ratio) and GFZ (floor-space ratio) 0.6; multi-storey houses were limited to GRZ 0.4 and GFZ 1.0 (Stadt Stuttgart 1980).

2. The second plan (Sta 91), passed in May 1983, solely added the area excluded from the former plan (Sta 70) as this had required further air-pollution tests on the project area. This was necessary because of changes in the above-mentioned federal environmental law (BImSchG). As a result the emission-protection area along the B10 had to be extended to between 60 and 80 m to the north and the protection wall itself had to be raised by about 6 m. The GRZ was determined at 0.4 and GFZ at 0.6.

3. The current local plan, the Stammheim-Süd II (Sta 95), approved in 1986, consists of the contents and areas covered by both previous plans. Additional changes concerned the height and width of the protection wall: a decrease of 6 m from the previous height. The changed building proportions: a GRZ of 0.3 and GFZ of 0.7, enabled an efficient density of buildings. These changes were heavily influenced by the contractors. Buildings must be constructed with sloping roofs in order to harmonize with the character of the neighbouring residential areas in Stammheim. This will be further ensured by restrictions on materials used and their colours. In general, the local plan contains many details and restrictions (Map 10).

The establishment of the local plan enabled housing development to take place despite the difficulties caused by noise emissions and limited building plots. Nevertheless the area now provides good living conditions, especially for families.

Reorganization of land holdings   In Stammheim-Süd the legal decision to undertake the replotting procedure was taken by the city council in 1970 and further activities did not occur before 1978. By that time, public participation of the citizens within the local plan procedure had taken place in 1977 and the intended development of the area had been put in more precise terms. The causes of this long time gap are explained above. The replotting procedure included the same territory as described within the local plan. The valuation of the land was made in 1980 and the replotting plan (Umlegungsplan) came into force in 1983 with a few exemptions in the northeast depending on the lodging of appeals by some landowners. They were settled in 1985.

In Stuttgart, *Flächenumlegung* (replotting according to space) is more common and this type of replotting was applied in Stammheim-Süd. But there are some important distinctive features to be noted, caused by the difficult conditions for developing a residential area at this location next to the B10.

One of the advantages of the replotting procedure is the early and complete extraction of areas reserved for streets, public greens and noise protection. These areas usually come to 15–25% of the total area, depending on the intended use and structure of the area. In the case of Stammheim-Süd, however, they came to 45% because of the noise protection wall.

The Bundesbaugesetz (BBauG – building code), which was in force until 1987, determined that the land-used for local streets and public greens (*örtliche Verkehrs- und Grünflächen*) had to be paid for by the municipality if it exceeded 30% of the replotting area. So in Stammheim the municipality provided the missing 15%, equivalent to 40,000 m², out of its own land. If the city of Stuttgart had not had enough land in its ownership, it would have bought it from private owners. That would certainly have delayed the development and the land would have been significantly more expensive. In 1987 when the Baugesetzbuch (BauGB) came into force, the law changed on this point. Since then, if the land values in the replotting area increase by more than 30% because of the development supported by the replotting procedure, exclusion of more than 30% of the area for local streets and public green space is allowed (Dieterich 1990: Rn 220).

The betterment levy within a replotting procedure depends on the increase in land values. In Stammheim-Süd the value increase resulting from the replotting procedure (*Umlegungsvorteil*) was great, coming to 90% (value in 1970 of the undeveloped plots before the procedure: DM110 per m²; value in 1970 of the developed plots: DM210 per m²); this corresponds to 47% in land (*Flächenbeitrag*). The municipality is allowed to retain a betterment levy of up to a Flächenbeitrag of only 30%, reduced by the percentage of land-used for the local streets and public green space. Thus, in the case of Stammheim-Süd, the city of Stuttgart could not take advantage of the betterment levy and from "gaining land", although there was a large increase in land values. However this apparent advantage of the private landowners was reduced by very high development fees.

The replotting procedure was necessary although the municipality owned more than 60% of the land, because the Realteilung had heavily fragmented the landholdings. A development based only on private instruments to change plots would have taken much more time. But in such difficult developments the instrument of replotting is not sufficient to solve all problems. The land

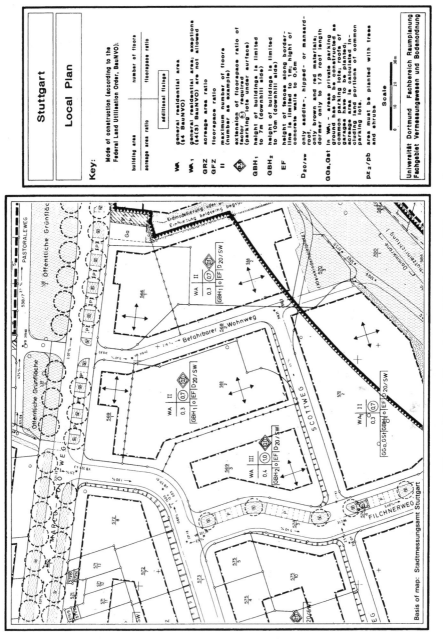

Stuttgart

Local Plan

Key:

Mode of construction (according to the Federal Land Utilisation Order, BauNVO):

building area	number of floors
acreage area ratio	floorspace ratio

additional fixings

WA   general residential area (§4 BauNVO)

WA₁   general residential area; exceptions (§4(3) BauNVO) are not allowed

GRZ   acreage area ratio

GFZ   floorspace ratio

II   maximum number of floors (number as example)

(25)   extension of floorspace ratio of factor required (parking lots under surface)

GBH₁   height of buildings is limited to 7m (downhill side)

GBH₂   height of buildings is limited to 10m (downhill side)

EF   height of fences along border-line is limited to 1m; height of concrete walls to 0,5m

D₂₀/SW   only saddle-, hipped- or mansard-roof, dormer only to 1/3 roof length only brown and red materials;

GGa,Gst   In WA₁ - areas private parking ground has to be constructed as common parking lots, roofs of garages have to be planted; acreage area is calculated including land portions of common parking lots.

pz₂/pb   area must be planted with trees and shrubs.

Scale
0   10   20   30m

Universität Dortmund   Fachbereich Raumplanung
Fachgebiet Vermessungswesen und Bodenordnung

Basis of map: Stadtmessungsamt Stuttgart

Map 10 Stuttgart-Stammheim: detail from the Bebauungsplan.

assembly of the city was of great importance for the development of this problematic area.

In Stuttgart the share of the land for servicing facilities increased in the 1980s decade and today amounts to 25–30%, because the land-take for public green areas increases and very often additional space for noise and pollution protection is necessary. This means that the betterment levy decreases today within the Flächenumlegung because of the legal limit of 30% Flächen-beitrag.

Installation of public utilities In general the costs for development facilities consist of three parts (independent of a replotting procedure):
○ the costs for the land value of the amenity areas (streets, public green space, noise protection, etc.);
○ the costs for surveying and the services of solicitors (the smallest part); and
○ the construction costs of the amenities themselves.

If the replotting instrument is applied, the first and second part of the costs mentioned above are part of the procedure and are involved in the value of the developed plots (*straßenlandbeitragsfreie Zuteilung*). In the case of Stammheim-Süd the provisions of development consist of the construction costs for all development facilities and one-third of the costs of the land. Thirty per cent of the area (two-thirds of the ground space of the develop-ment facilities) are free of charge within the replotting procedure. The additional 15% provided by the municipality (one-third of the ground space) must be paid. The latter part of the costs is included in the Umlegungsvorteil and has not to be paid in this case.

The improvement costs (*Erschließungskosten*) in 1987 totalled DM19.96 million. The following items are included:
○ DM5.85 million for streets, pavements and lighting;
○ DM0.30 million for pedestrian routes;
○ DM1.19 million for public green space;
○ DM0.38 million for noise and pollution protection; and
○ DM11.90 million substitutional payment for the additional 15% of land provided by the city.

Of this, 10% had to be paid for by the municipality, while DM18 million was divided between the landowners. This means that DM170 per m² of permissible floorspace area (GFZ) was the development fee for each landowner, even when floorspace area was not utilized to the maximum. When the density was restricted to a GFZ of 0.7 in the local plan, DM120 per m² of the plot area had to be paid.

In most cases the development fees were charged by so-called Ablösever-träge (discharge contracts). The intention of these contracts is that landowners pay a fixed development fee before the public utilities have been constructed. Landowners receive financial benefits if required costs for the services provision are above the fixed development fee. Finally, the real costs were determined in 1987 after completion of all construction in the area, which had already begun three years earlier.

With regard to the largest expenditure, all landowners, including the city, had to pay for the 15% additional land. DM11.9 million was fixed by the city in 1984, which valued the price of the required land area at DM300 per m². The value is based on the 1983 situation, when developed land in this area was valued at approximately DM600 per m². It is expected that the landowners, who did not conclude discharge contracts (Ablöseverträge), will dispute these unusually expensive development fees in the courts, especially because of the fixing of the value of the land at DM300 per m².

By comparison, the construction costs of the wall were unusually low. The municipality used the wall as a deposit area for building waste. It is common to raise fees for depositing material. So the main part of the construction costs could be refinanced.

The actors and their behaviour
Besides the local government, other actors, especially building contractors, companies and private individuals, played a rôle in the process of housing development.

The city's housing programme Because of the level of land prices, even individuals with moderate incomes find it difficult to acquire property under open-market conditions. At the most, the city is able to subsidize high land prices in specific locations. A special housing programme has therefore been established by the city government known as *Preisgünstiges Wohneigentum in Stuttgart-Stammheim-Süd*. Under this programme, which applies only to city-owned plots, families from the middle class and people with low incomes are encouraged to acquire housing property at favourable costs.

The programme was implemented in three phases. The first phase, established in 1984, contained 87 family-owned houses and 37 apartments. Construction was finished in 1985 and, because of the increased demand and its great success, a second phase was continued in 1985 with a development of 110 family-owned houses. The buildings provide living space of about 111–121 m² on building plots of 197 m². Several conditions were taken into account by the contractors:
○ The price for the houses in the latter phases, including all finishing, was limited to DM264,000, with the price for the building plot excluded.
○ The development concept targeted the construction of solid but low-cost family-owned houses, which were to be built to a moderate standard but

with the potential for future quality extensions.

○ Materials and construction work at favourable costs were preferred. In order to reach these conceptual commitments, the type of building was restricted to terraced houses.

Potential home owners had to meet several conditions to apply for one of the housing units:

○ The basic criteria is set with the level of household income as determined by §25 II. WoBauG (legislation on house building) plus 40%. This equates, for example, to a three-person household with an annual income of about DM55,500.

○ Equal importance is given to the status of the Stuttgart citizens. Single-parent families, young families and those with three or more children already living in the city are given preference. Exceptions can be made for applicants working but not living in Stuttgart.

○ The applicants are not allowed to own building land, residential buildings or apartments.

○ The required equity ratio was set at between 15% and 20% of total costs.

After the replotting plan was completely validated, the offices for both housing and municipal property were responsible for distribution of the building plots according to the Stuttgart housing programme.

The municipality subsidized the land price of building plots affected by the programme. Two different kinds of land-price financing were available to the new home-owners, but in both cases the land owners had to pay DM120 per m^2 development fees:

○ First, to purchase the plot immediately at a subsidized price fixed at DM250 per m^2, which was 60% less than the open market value at that time of DM625 per m^2.

○ Secondly, to acquire an hereditary leasehold, the ground rent is subsidized and determined at 4% out of DM250 per m^2 per year, that is DM2,000 per year. The land must be bought after 10 years at the latest, otherwise the hereditary right is cancelled.

○ Additional funding was provided by the federal programme, known as *kosten- und flächensparendes Bauen* (cost-and-layout-efficient buildings) with loans of DM32,000 to each home-owner. It was included only in the first phase of the Stuttgart housing programme.

○ Low-interest loans of DM180,000 per household, equivalent to the *II. Förderweg*, dependent on current *Landeswohnungsbauprogramm* rules provided by the state government, were available for all households involved in the Stuttgart housing programme.

In 1987 construction of the first and the second phase was completed. In 1986 the third phase of the programme began, providing an additional 81 family-owned houses. In the first phase, development took place in the western part of the area and was continued in the next two phases in the eastern part. Provision of services was implemented at once in the whole area. In 1989 the programme was extended to two other city districts of Stuttgart.

The Stuttgart housing programme is valued positively by both the city and its citizens. The city is doing its duty to allocate cheap land for housing (§89 II.WoBauG) with great success, and families have the chance to acquire their own property at favourable prices. By establishing the Stuttgart housing programme, the level of land prices was subsidized from DM800 per m² under open market conditions to DM250 (1984 values). For the new home-owners in the mid-1980s the subsidies resulted in savings of DM75,000 per 200 m² plot or savings of DM3,000 per year on the ground rent. This is another example of the importance of land banking by the municipality.

From the standpoint of local government the whole of the difference between open-market value and actual land prices achieved cannot be seen as subsidy. The expenditure is influenced by the prices for undeveloped land in the 1960s and 1970s and the financing costs of approximately 15 years. The amount the municipality will have invested in Stuttgart-Stammheim for land banking may be estimated to be between DM14 million and DM18 million. Increased values because of development of the area favour the municipality. Whether expenditures and revenues are in balance is unknown, but it is to be expected that after reckoning up the balance would be negative for the municipality. This is mainly caused by the financing conditions of land assembly.

The programme illustrates a part of land policy implemented by local government that applies to middle-class citizens. Indeed lower-class households are not represented, because land prices and required subsidies seem to be too high to allow these groups to acquire property. It is questionable whether it would make sense to support the underprivileged to acquire land property in a region with such high price levels as in Stuttgart.

Building contractors In Stammheim four main contractors were involved: Gemeinnützige Baugesellschaft Zuffenhausen (GBZ); Wüstenrot, Landesentwicklungsgesellschaft (LEG) and Neue Heimat (NH). These are private associations under the former Gemeinnützigkeitsgesetz, which set restrictions in terms of profit until 1990, the year when it was cancelled. The first one, the GBZ, is more regionally orientated and the other ones act at the national level.

There are many small contractors in Stuttgart, most of whom have specialized offices with advantageous connections to local politicians. Traditionally they limit their activities to single city districts. The most important of them is SWSG (Stuttgarter Wohnungs- und Siedlungsgesellschaft), a public association wholly owned by the municipality. None of these was involved in the housing programme.

The contractors had to apply to the local authority for work, with preliminary calculation and drafts of different types of building designs included. With regard to the housing programme, the contractors were responsible for the development of a specific portion of the sites owned by the city after they had been chosen by the local authority. In agreement with the government and the construction companies, each contractor concluded development designs for a specific type of building to be constructed under the housing programme regulations. They also were obliged by the city to set fixed prices for housing construction to a certain quality standard, although individual details such as car ports, were excluded. To bring about development, the main contractors had to subcontract tasks to various smaller construction companies.

All the four contractors were employed by the city in order to keep them in competition, which resulted in acceptable and trouble-free housing development. Even with regard to the fixed building costs as determined by the city, the contractors were able to make a profit within the legal contract restrictions. This depended heavily on using standardized components, on combining orders for similar construction works at different locations, and on relatively simple types of buildings with little scope for individual construction details. Here intensive compliance with contracted construction companies was required. In addition, several home-owners ordered individual construction details, not subsidized by the housing programme but offered by the contractors.

In summary, the main reason for the contractors to get involved in the housing programme was to secure their own employment situation and revenue flow. Because of the housing demand in Stammheim, a commercially successful outcome was never in doubt.

Construction companies It is not usual to contract an ordinary company for the construction of a large area of building land. However, the contractors mentioned above were not allowed to operate as construction companies, so they employed various local small and medium-size companies as contractors for the erection of the houses according to the architects' plans.

Subcontractors employed by the main contractors were responsible for the actual physical construction. They had to ensure the meeting of fixed prices for the buildings as described above. Here the most important solutions were reducing the thickness of dividing walls and using a specific type of ceiling. These construction details enabled building costs to be reduced significantly, but this was disadvantageous for noise reduction.

Home-owners not subsidized by the housing programme also employed construction companies to build their own houses. Thus completion of houses in a short period of time was ensured. It is important for most of the home-owners to act under open-market conditions in order to minimize the period of interim financing of unused building plots.

Home owners Applicants chosen by the city under the housing programme had to choose a building type as provided by one of the four contractors. Accordingly, the contractors had no problems selling their houses, although there was competition between the four contractors. At least they were able to refuse those applicants with significant financial problems, and applicants had no legal claim.

Home-owners in Stammheim, who purchased their property under the conditions of the housing programme, generally preferred the hereditary leasehold. Building land is transferred to the home-owner at a ground rent far below the market value, and interim financing under favourable conditions was available.

It is also possible to extend the set deadline of 10 years, following a decision by the city council, if home-owners cannot purchase the plot from the city within that time. Nevertheless most of the home-owners purchased the building plots after a few years. On the one hand prices for built-up land increased continuously and on the other they only had to pay 60% of the current market value for building land.

Most of the free-standing one- to three-family houses, which have been erected within the project area were financed by private persons who were not participants in the housing programme. In many cases the houses include apartments that have been rented by the house-owner to other households on the open market. Here prices are at maximum levels.

Ownership of land is common in the region of Württemberg as a result of Realteilung. Inheritance or owner-occupation of an apartment is the basis for financing private housing. Accordingly, new home-owners typically finance private or public housing by ownership of land or buildings that have been sold previously. As a consequence, the process of replotting is usually a prerequisite of the common pattern of housing finance described above.

7.3 Dortmund

The Technologiepark in Dortmund is a case of a greenfield project that was successful because of intensive public-sector involvement and subsidies. In Germany's industrial sector the project represents the more usual market conditions, in contrast to the investment market conditions described in the Düsseldorf case-study (Ch. 11.1). Various private and public actors such as the Land, the regional government, the municipality and the chamber of industry and commerce achieved their aim of attracting high-quality industry into a structurally weak region. In such places – peripheral or old industrialized regions and marginal zones of agglomerations – free-market conditions do not exist on the industrial land market.

In these regions the heavy influence of the public sector on the industrial land market often creates an artificial land price. Because of the competition between the municipalities, the prices on offer are too low and therefore the municipalities are deprived of financial benefits, which would be possible under free-market conditions. The price of the real estate has hardly any influence on investors' decisions. Furthermore, there is no government control of land prices and rents for industrial buildings.

The municipality is responsible for the development of industrial areas within its boundaries. The cost of developing and servicing the plots has to be borne initially by the municipality. Later, the owners have to pay 90% of all costs for servicing their plots, but there are special subsidies available for this purpose from the Bundesländer and the Bund. Usually public spaces (such as streets and green spaces) stay in the ownership of the municipality.

In Germany's industrial market it is very common for a single entrepreneur or company to manage the acquisition of land as well as the construction of the buildings which it will occupy itself. Rental and leasing of property is not typical, although this has increased in the past few years. Used premises are normally sold as a unit including the land. The Erbbaurecht, or hereditary leasehold, is also applied. The creation of buildings in stock by the municipality is not common in Germany, although this kind of industrial property market has recently increased.

Potential investors are normally informed about land and property prices directly by the municipalities. It is not very common to hire a broker or an estate agency. Only second-hand premises are traded by private individuals. Here the private owners usually advertise their properties through the newspapers.

In recent years the *Wirtschaftsförderungsgesellschaften* (mixed economic assistance agencies) have been formed as new bodies involved in the property market, especially in the large cities. They have the same aims as the Wirtschaftsförderungsamt (economic assistance board), but are more flexible and are less involved in the communal policy.

The geographical and economic situation Economic development in the past, the current situation and future economic development all have to be taken into account when considering the Ruhrgebiet industrial area. The region has about 6 million inhabitants and can be regarded as one of the largest industrial zones in Europe. Dortmund is located in the east and belongs to the EC Region Number D. 55 Region Arnsberg. The economic structure of this area had been formed during the industrial revolution, so primary industries such as coal-mining and steel production were dominant in the past. A high population density and a disordered spatial structure have caused an unfavourable image for new investors.

Economic growth comparable to that in other older industrialized areas in Europe or the USA has taken place especially in the southern area of the Ruhrgebiet, known as the Hellweg Zone. This includes the main cities of Bochum, Essen and Dortmund. Meanwhile, the economic situation has stabilized, although in the northern area of the Ruhrgebiet, the Emscherzone, economic growth is much slower. Serious problems such as high unemployment and derelict land remain to be solved. Economic growth has occurred

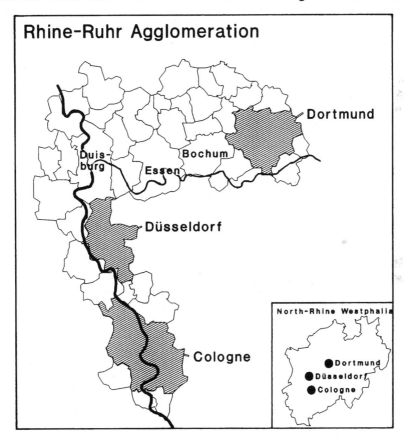

Map 11 Rhine–Ruhr agglomeration.

because of effective co-operation between the local and the regional government and the private sector. Their main aim was to improve the "economic atmosphere" for new investors.

The public authorities, including the Land, set up several new financial programmes and combined existing grants for the region. Most of the money has gone towards infrastructure improvements, renewal projects, a special land-banking system of derelict land and as direct subsidies for companies.

In spite of its problems the Ruhrgebiet is well situated in the middle of Europe, has excellent transport links and offers well educated and skilled workers.

Dortmund itself has about 600,000 inhabitants and covers an area of 280 km^2. Dortmund is an Oberzentrum, or central city, with a surrounding population of about 2 million. To the north, south and east it is surrounded by large peripheral recreation areas. Although the traditional steel and beer industries still remain, new companies, especially those in the service sector are settling in the Ruhrgebiet (Hennings & Einem 1988). For example, Dortmund has became one of the largest centres of German insurance companies. In 1989, 62% of workers were employed in the service sector, compared with 50% in 1977. There are currently about 26,500 companies, employing about 240,000 people (Wirtschaftsförderung Dortmund 1989). Since 1968, Dortmund has had a technologically orientated university with some 25,000 students. With several motorways, a railway and a canal port, the transport links are excellent.

The industrial land market and land policy

The industrial land market in both the Ruhrgebiet and in Dortmund is predominantly determined by supply. This is mainly promoted by an intensive land-supply and land-banking policy on the part of the municipalities. In general, in the cities of the region, the present supply is sufficient, although there is a deficiency of prepared land in well situated locations, such as on greenfield sites.

Prices for industrial land and for industrial properties are much lower than in southern Germany. There is less pressure, owing to a surplus of demand in cities such as München or Frankfurt.

Most municipalities face the dilemma that they are providing mainly derelict but well prepared land (industrial brownfield land), while companies are demanding greenfield sites, which have a better image. Conflicts over use of ecologically important areas constantly arise. In addition, there is a low level of acceptance by the public of the use of greenfield sites for industrial development. There are also increasing conflicts with the Regionalplanung (the regional planning authority), which attempts to prevent excessive greenfield development.

The difficulties associated with the provision of formerly used or contaminated industrial land and in attracting new firms to it is exacerbated by the owner-occupiers. In the past the coal-mining and steel companies were not interested in selling their plots. Now they are trying to commercialize their industrial sites, or parts of them, by themselves, and problematical plots of land remain.

In Dortmund the current market situation of supply and demand is mainly influenced by the activities of the municipality. The planning department is responsible for providing sufficient new industrial land within the framework of the Flächennutzungsplan, the preparatory land-use plan, which includes about 1,700 ha of used and undeveloped industrial land. Because of the large steel companies and former coal mines, only about 830 ha of land is used for other industrial purposes or for office space (see Map 12). Next to the planning department is the Dortmunder Wirtschaftsförderung (the economic assistance board) which belongs to the administration of the city and has become more important in recent years. The board is responsible for the acquisition of new firms, for offering suitable plots, for promoting the city, and also provides a full service in the form of consultations, concerning, for example, available subsidies or planning permission requirements.

Only 250,000 m^2 of reserves of Bauland are available in Dortmund. 610,000 m^2 of greenfield land are available, but plots of over 30,000 m^2 are not available. In the north of Dortmund there are about 250 ha of brownfield land which is not available because of industrial pollution or because the owner-occupiers (mainly the coal-mining companies) are not interested in selling the plots (see Map 12).

In recent years, however, the Grundstücksfond Ruhr, which manages the buying, decontamination, redevelopment and sale of derelict land, prepared 80 ha of brownfield land. Most of this land remains unused because of its poor image and the low demand for this type of industrial land.

The prices in 1990 for prepared industrial land varied between DM40 and DM130 per m^2 (Valuation Committee Dortmund 1990). By comparison with other large cities in Germany this is very low: for example, in München prices of up to DM1,000 per m^2 are typical. Only in a few prime locations (for example for office space along the motorway B1) can land prices reach DM180 per m^2 (see Map 12). In Dortmund prices for renting industrial space vary between DM4 and DM8 per m^2 for production and service space. The prices for office space do not exceed DM22 per m^2.

The main public policies concerning the industrial land market can be summarized as follows:

○ an increase in the number of staff at the economic assistance board;
○ generation of new initiatives to support industry and commerce;
○ promotion of the advantageous location factors;
○ provision of a sufficient supply of prepared industrial land;
○ renewal of derelict land;
○ encouragement of investment by new firms;

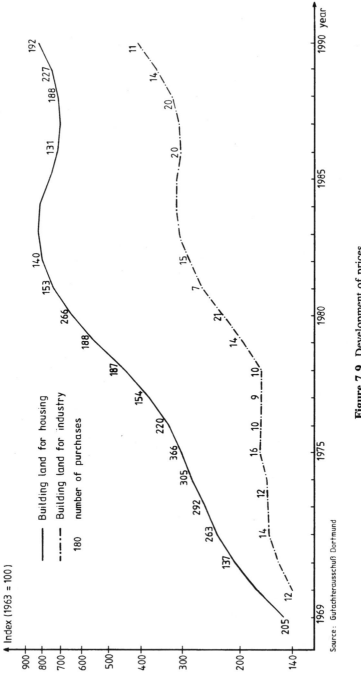

Figure 7.9 Development of prices.

○ assistance by speeding up the process of gaining planning permission;
○ advice concerning available local, regional, national and European financial programmes and;
○ supply of information on land prices and development costs.

Description of the project

The Technology Park was located on a greenfield site owned by the city of
Dortmund (Map 12). The site was originally allocated for the expansion of
the university in 1984 in the Gebietsentwicklungsplan (regional plan) and
also in the city's Flächennutzungsplan. The site borders one of the few
remaining and therefore important large clean-air corridors of Dortmund,
which are protected by the Regionalplan. These are corridors to the inner-
city area. Conflicts continue between the city and the regional government

Map 12 Business location.

179

over the future use of the site, because of the intended enlargement of the Park.

Currently the Technology Park is still in a phase of growth. After its final completion it will cover about 19.5 ha and house between 80 and 100 firms employing about 4,000–5,000 people. The Park mainly consists of a mixture of technology orientated, medium-size firms such as those involved in microchip production, material technology, environmental technology and research & development. Only half of the site is involved in production; the rest consists of office development.

The firms gain many benefits from the exchange of staff, research results and development techniques provided by the University of Dortmund. The companies that have settled here are aware of the benefits of this location. Beside the potential of well educated employees from the technology-orientated university, other benefits include the environmental landscape and the favourable traffic connections, such as a direct motorway connection to the centre of the city of Dortmund.

The origins of the Technology Park was combined with the Dortmunder Technology Centre which is located next to the Park. The Park and Centre had been planned together, although the construction of the Centre began first. It was the idea of the chamber of industry and commerce and the local economic assistance board to utilize the innovative potential of the university for commercial application. The project was financed by regional subvention programmes of the Land of Nordrhein–Westfalen, the city of Dortmund, the banks of Dortmund and the chamber of industry and commerce.

The Technology Centre offers opportunities for new and growing technology-orientated companies and facilitates a relationship between university and industry that is necessary to encourage economic growth. Companies rent space for a limited period of between three and four years. The prices vary between DM16.5 for office space and DM14 for production space. After this period the companies should be more financially stable and able to move to the Technology Park.

The design of the Technology Park attempted to create an architecturally significant setting, which differs clearly from the appearance of conventional industrial parks. Particular importance was given to the high quality requirements of public areas, of building fronts, of the heights of the buildings, of the parking space and of the green areas (Map 13). In order to ensure these architectural requirements the local authority utilizes the restrictive and detailed Bebauungsplan and a *Gestaltungshandbuch* (manual for design criteria). This allows potential investors to obtain detailed information about the construction opportunities and costs.

The layout design of the Park follows a strict square-matrix pattern that forms single block structures. These blocks consist of so-called site modules

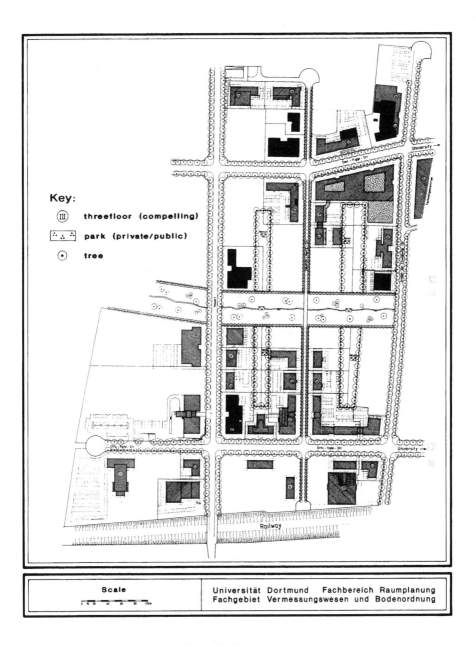

Key:

(Ⅲ)   threefloor (compelling)

⸬   park (private/public)

⊙   tree

Scale

Universität Dortmund   Fachbereich Raumplanung
Fachgebiet Vermessungswesen und Bodenordnung

Map 13 Site layout.

with areas of between 850 and 1,900 m² (Fig. 7.10). The offices and the other service buildings that front onto the streets have to be constructed with three floors, along a fixed *Baulinie* (building line). Dominant materials (particularly red brick, steel, glass) must be maintained throughout all buildings.

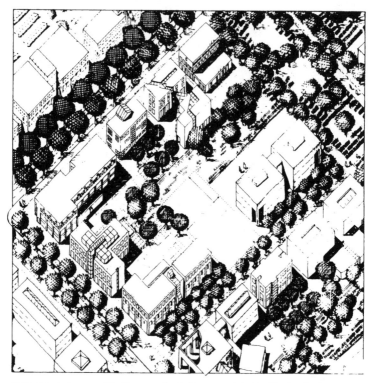

Figure 7.10 Modulsystem II. (Source: Stadt Dortmund (ed.), Technologie Park Dortmund, Documentation No. 4, Dortmund 1990: 21).

The land is sold at a price of DM130 per m². This comprises DM70 for the cost of the land itself (according to the *Richtwert* or proposed land value), DM30 for development fees and DM30 for the first planting.

The planning costs totalled about DM1.7 million for the whole area surrounding the university, among these DM440,000 for the Technology Park, DM320,000 for the landscape planning and DM580,000 for traffic planning. The whole planning costs came from resources of the municipality and from the Ministry for Town and Country Planning, Housing and Traffic in Nordrhein–Westfalen (MSWV). The public investment for the new developments and for new landscape design reached a total of about DM15.5 million. The development of the area was carried out by the municipality and cost about DM30 million. Roads and sewers remain in the possession of the municipality. Up to now, the private sector has made an investment of approximately DM150 million, so the ratio between public and private investment is of the order of one to ten. Figure 7.11 illustrates the different stages in the planning and implementation process.

Project implementation

The municipality – aims and instruments The planning and development of the area and the attraction of firms to the Park was carried out completely by the municipality. It was the result of cooperation between various departments in the local authority (town-planning, economic assistance, green-area office, environment office) and a leading group made up of representatives of the state and district government, politicians from the city council as well as representatives of the chambers of commerce and a consortium of banks. In addition, a project group was formed. It consisted of specialists of the planning department, the economic assistance board and private architects, which have been commissioned by the city. The group worked under the leadership of the Oberstadtdirektor (the director of the city administration). It prepared the plans for the Park, which were discussed at length with the leading group.

The city of Dortmund adopted a market policy of extensive supply. As in similar cases elsewhere, the local authority became a Zwischenerwerber, or intermediate acquirer of land, by buying the land from the former owner: in this case from the Land of Nordrhein–Westfalen. The Land sold the area at a very low price. The municipality developed the land (replotting or compulsory purchase were not necessary) and offered the newly developed building plots to interested companies through nationwide newspaper advertisements. The promotion of industrial land and the attraction of new firms has become more the responsibility of the municipality's economic assistance board.

In addition to the aim of supplying building plots, the municipality was interested in the orderly development of the university surroundings. The city of Dortmund had already recognized the development opportunities of the site located next to the university and its rôle in encouraging economic growth. Despite the fact that several companies intended to move onto the site as soon as possible, the city council had the foresight to stop the establishment of the potential investors for two years until an integrated planning concept was prepared (Günther 1988: 1556).

An essential part of the planning process was a *Bereichsentwicklungsplan* (special area development plan) for the area around the university and the preparation of the Bebauungsplan.

The Bereichsentwicklungsplan determined the framework for the later local plan of the Technology Park. The plan illustrates the future use intended for

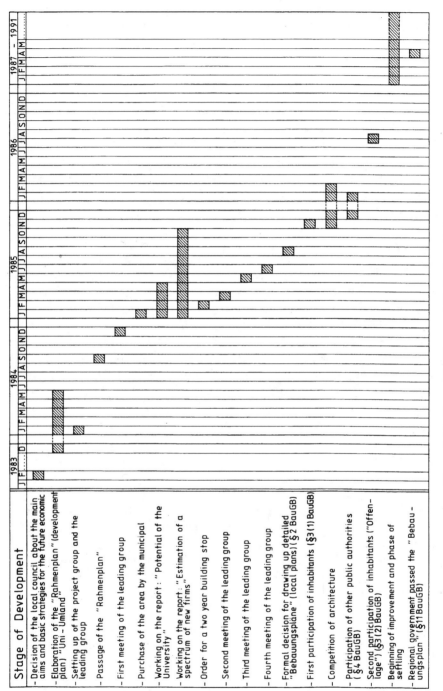

Figure 7.11 Time-table: Dortmund Technology Park.

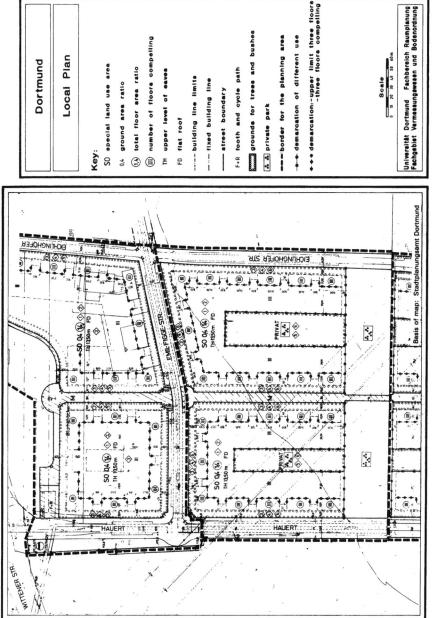

Map 14 Local plan (extract).

Basis of map: Stadtplanungsamt Dortmund

Dortmund

Local Plan

Key:

SO special land use area

0,4 ground area ratio

(1,6) total floor area ratio

(III) number of floors compelling

TH upper level of eaves

FD flat roof

---- building line limits

---- fixed building line

---- street boundary

F+R footh and cycle path

grounds for trees and bushes

private park

--- border for the planning area

--+-- demarcation of different use

--•-- demarcation: upper limit three floors
             -three floors compelling

Scale

0 10 20 30 40 50 60m

Universität Dortmund Fachbereich Raumplanung
Fachgebiet Vermessungswesen und Bodenordnung

different parts of the whole area and takes into account the linking of uses and the environmental protection aspects. The subsequent planning process for the Technology Park was mainly influenced by this plan, especially concerning the acceptance of the project by the regional planning authority and neighbouring citizens. The plan demonstrates measures such as landscape design and attempts to reduce the environmental impacts of such a greenfield site development.

The Bebauungsplan was the most important instrument for setting out the aims of the project (see Map 14). The requirements of the local plan were rather restrictive and somewhat atypical. The whole area is allocated as a *Sondergebiet* or *SO-Gebiet* (special district). This means that only advanced-technology-orientated firms from special branches such as those involved in the chemical, electronic or computer industries are permitted to locate there. Manufacturing industries, with the exception of research and development companies, are excluded.

The plan fixes the exact location of the buildings on each plot by special building lines. The buildings must be three storeys high to the street side and can be lower at the back. The buildings have to be built in red brick. Roof-shave to be flat. A special part of the back of each individual site has to be provided for a private car park. Each single plot of $800\,m^2$ also has to be planted with two trees with large canopies, three trees with medium canopies and 300 shrubs.

The ground-area ratio is 0.4, while the total floor area ratio is 1.4. In some individual cases, so long as the total floor-area ratio is observed, a ground-area ratio of up to 0.5 may be permitted. Exceptions are allowed if the loss of land is compensated for by additional tree and shrub planting. Every square metre of the building that exceeds the ground area ratio of 0.4 has to be compensated for by green space on the roof. The upper level of eaves should exceed $13.5\,m$.

Further stipulations concerning building design criteria are given in a Gestaltungshandbuch (manual for design criteria) that explains the aims of the intended urban design, provides recommendations for building design and illustrates the building obligations. Each company has to comply with it and the municipality controls the restrictions through the building permission procedure.

The purchase contracts are the other important aspect of the implementation of the project. The municipality, as the owner of the plots, instructs the interested buyer on the building design criteria and thus maintains considerable control over the development as intended by the local plan. The project can therefore be regarded as a very successful example of the interaction between the public planning and building law and the private law of purchases.

Special clauses in the purchase contract allow the following:

○ the municipality to buy back the land within the next 10 years if the buyer does not use the property in the agreed manner. The buy-back price is fixed at DM130 per m^2;
○ restricted leases for the municipality (beschränkt persönliche Dienstbarkeiten);
○ the beginning of construction within 4 months after signing the contract;
○ the termination of construction within 12 months;
○ responsibilities for the design of the buildings according to the manual for design criteria and for open areas according to the green plan; and
○ extensive rights for approvals of changes of land-use, resale, renting, and changes of design.

The limitations imposed for each company follow extensive discussions and detailed advice from well illustrated information brochures.

The companies There are two types of procedure for occupying the land. First, an interested firm buys the plot from the municipality and acquires a new owner-occupier status. If the firm intends to sell the plot it has to pass on all important requirements included in the first contract with the municipality. These are known as chain agreements.

Secondly, firms are able to rent or lease suitable floor space. In the beginning of the Park this was not common, but it has increased in the past few years. Newly established companies with limited company capital use this way of occupying the land. According to the purchase contract with the municipality, building contractors have the commitment to rent the space for a minimum period of 10 years. The tenants have also to be approved by the municipality and a cancellation of a renting contract has to be announced by the municipality.

Conclusions

With the consistent application of planning instruments, the creation of the infrastructure in advance, land banking policy, massive subsidies as financial aids, advice, and by intensive promotion, the municipality of Dortmund achieved its aim of attracting technologically orientated new firms to the Park. The municipality increased the supply of building land as a method of bringing about economic growth and a more stable economy in Dortmund. It was the interaction between the main public authorities, such as the municipality of Dortmund, representatives of the government of the Land of Nordrhein-Westfalen and important institutions of the regional economic community (for example, business associations, chambers of industry and commerce, and consortiums of regional banks), which have influenced the successful implementation of the project.

The purchase process of the land is a further example of public aims being achieved through the consistent use of public law, such as planning legislation, and private law for, for example, purchase contracts.

The Park illustrates the general trend towards higher levels of building and design quality of industrial developments. The Technology Park tries to point out the present and the future trends, and it is possible to transfer them to other areas in Germany.

This case study demonstrates both the advantages and disadvantages of the federal planning system. The municipality maintains a powerful rôle and has the most of the responsibility for planning, so that the city of Dortmund is able to influence the supply of industrial building land and to carry out a successful project. The arrangements with the regional planning authority must also be noted. It is clear that, in the main, ecologically important greenfield land is allocated for industrial uses, while several hectares of brownfield land in the north of Dortmund are still unused, reflecting the competition between neighbouring cities of the Ruhrgebiet.

The project shows the importance of the *Bereichsentwicklungsplanung* (development plan for special areas). It is a very flexible instrument because it is not binding on the municipality or private persons, and it significantly increases the public's acceptance of the development. The Flächennutzungs-plan is a suitable method to determine the supply, while a restrictive Bebau-ungsplan is able to guarantee security for the companies and it makes a high quality of building possible.

The traditional instruments are ineffectual (especially for implementation), if they are applied alone and not supplemented by additional "soft" instruments such as information, advice and promotion. Furthermore, a real-estate contract with strong limitations and binding agreements is very helpful if the municipality is the landowner. The Technology Park shows another important aspect of the industrial land market in stagnating agglomerations. Without massive financial intervention, such a success would not have been possible.

PART III
The urban property market

CHAPTER 8

The framework of
the urban property market

In Germany the land market and property market systems are indistinguishable. The most important aspects of the framework of the property market are the same as those of the land market, already described in Part II, although there are specific regulations covering the property market that are important to its functioning. In the German property market, a separation between the new and second-hand market (that is between the primary use and the secondary use of premises) is in general not relevant.

8.1 The legal environment

Legislation and the legal authorities relating to the urban property market are generally the same as those relating to the urban land market. The statutes and planning legislation relating to the land market cover aspects of the property market. For example, the most important regulations in private law are the regulations of the *Bürgerliches Gesetzbuch* (BGB), which include legislation concerning renting that is especially important in the industrial and office sectors. On the property market the most important statute within public law is the Baugesetzbuch (BauGB), followed by the Baunutzungsverordnung (BauNVO) and the building regulations of the Länder.

The environmental protection law is relevant, particularly in the industrial sector. The Bundesimmissionsschutzgesetz (BImSchG), for example, influences the property market in areas that include both industrial and residential uses (*Gemengelagen*). In the housing sector, noise protection often influences property values and their standing on the property market.

In general the legal principle of *Bestandsschutz* is valid. This means that a building or land-use already in existence must not be changed or demolished if it would be illegal under current law. It retains its legal status if it was legally established under former law. This is important on the property market, especially because of the increasing requirements of environmental protection, for example within Gemengelagen. Nevertheless changes of use or modifications to buildings have to meet the requirements of current law. In the housing sector, rent controls, including those applied to social housing, are of special significance. For example, federal legislation states that rents have to be based on those charged in similar locations and for similar flats in the previous three years. Rent increases are normally limited to 30% over the following three years, unless building improvements are carried out. Another aspect of the legislation is protection against wrongful eviction and exorbitant rents. There are only a few acceptable reasons for a landlord to quit a flat (§564b BGB).

One special factor relevant only to the property market is the listing of buildings and monuments (*Denkmalschutz*). The listing of buildings is regulated by legislation at the Länder tier of government. The counties and large municipalities have jurisdiction over such listings, which may be made without the agreement of the owner. In special cases it may be necessary for the municipality to take over the property if the owner is unable to pay for the upkeep of the building. It is also possible to designate a group of buildings as a conservation area (*Denkmalbereich*). The listing of a property leads to significant restrictions on modifications to appearance and on changes of use, but it also provides some tax advantages and grants. In view of the value of a property, listing is sometimes considered a disadvantage (Kleiber et al. 1991: Rn 1020/1210), but if such a building is one of a group of historic buildings, or is used for a special purpose (such as a restaurant) an increase in value is to be expected.

Possibilities of public intervention

Technical conversion and change of use A technical conversion of a property, change of use or demolition requires official approval. In effect, this approval amounts to new planning permission. There are only a few exceptions to the rule. The legal base for technical conversions is given in Länder building regulations that also cover the demolition of premises. The Länder building regulations (*Landesbauordnungen*) as well as the BauGB embody regulations for changes of use. Without approval, an owner may not change a flat into an office or shop; nor may premises within an industrial area be converted into flats. Approval is required even if there would be no change in the appearance of the building.

A specific instrument to prevent changing residential living space into other more profitable uses is given to the Länder in federal legislation (*Zweckentfremdungsverbot*). Thus the Länder are empowered to instruct specified municipalities to grant special approval in the case of change of residential use to another use. This instrument should protect housing space in areas with very scarce housing markets. It is used by many of the Länder, especially in the metropolitan areas (see the Frankfurt case study, Ch. 11.2).

Preservation of town districts In order to reach the following three goals, a municipality may implement the instrument of the *Erhaltungssatzung* (§172 BauGB):

o the preservation of the character of a district or a smaller area with regard to the appearance of the buildings. It may only be used for valuable building stock that is not designated as listed buildings;

o the preservation of a mixed population structure within a district or a smaller area and;

o the reduction of negative social effects during the redevelopment of an area.

A city council may set the boundaries of a district or area by local law. Within the area of the Erhaltungssatzung, a municipality has the right to reject applications for the demolition or modification of buildings and change of use for the reasons mentioned above. The first goal is concerned with the quality of a residential area, while in the second and third cases social objectives are paramount. The municipalities attempt to use the Erhaltungssatzung to prevent flats being converted into apartments in attractive districts constructed at the beginning of the century (see the Köln case study, Ch. 11.3). It has become apparent, however, that the instrument is not very effective in preventing such development.

Renewal and redevelopment There is a procedure set out in law for the redevelopment of an urban area (*Sanierung*, §136–164 BauGB). This instrument can be used by a municipality if there are shortcomings and problems with the existing development of a district. This can refer to both residential areas and to industrial areas such as abandoned or derelict former industrial locations. The redevelopment procedure is carried out by the larger municipalities themselves or by specially commissioned companies (*Sanierungsträger*).

Since 1971 the procedure has been based on specific legislation: the Städtebauförderungsgesetz. In 1987 it was integrated into the Baugesetzbuch (BauGB). There is the normal extensive redevelopment procedure and a simplified procedure, which was introduced in 1985, because normal Sanierung proved too inflexible in many cases.

The property owners within a redevelopment area have to accept even more restrictions than in a replotting procedure. Almost every modification of the property and of its use requires official approval. This includes contracts for intended sales and acquisitions or for the use of parts of a property. In agreement with the owner, the municipality has even extended powers to terminate a tenancy with compensation (§182ff BauGB).

The improvement of public residential amenities in the area is carried out by the municipality. This normally includes the reorganization of landholdings according to the new pattern of use given by the local plan, often done by the replotting procedure. It also includes new streets and public utilities. These actions are jointly financed by the Bund, the Länder and the municipalities. Normally each tier provides one-third of the costs. The improvements of the private properties are managed by the owners. Many subsidies, as well as tax incentives, stimulate the owners to improve their properties.

In redevelopment areas, property values increase as a result of the public intervention described above. Within normal redevelopment procedures the municipality is allowed to levy a charge for betterment. At most this may be the difference in value before and after the renewal procedure. Arriving at the pre-renewal valuation presents difficulties, because it should not be influenced by speculative values and the anticipated effects of the intended redevelopment. The municipalities seem to regard betterment levies as problematic and they are seldom used.

Renewal of districts is often implemented without these legal instruments, in which case there are no restrictions for the owners. Improvements of public residential amenities, such as traffic calming (Verkehrsberuhigung) and living conditions in a district or block (*Wohnumfeldverbesserung*), are financed by the municipalities and receive substantial grant aid from the Länder, but not from the Bund. Investment by private owners is stimulated in the same way as described above. Sometimes contributions based on the Kommunalabgabengesetz (legislation for local fees and contribution) are charged by the municipalities.

§175–179 BauGB enable municipalities to require a property-owner (or the municipality at the owner's expense):

○ to carry out development on it (*Baugebot*);
○ to modernize or maintain buildings if their condition is becoming dangerous (*Modernisierungs- und Instandsetzungsgebot*);
○ to plant trees and shrubs according to the regulations of a legally binding local plan (*Pflanzgebot*);
○ to demolish a building if it is incompatible with the local plan or it has fallen into disrepair (*Abbruchgebot*).

In each of the cases it is possible that some of the costs are borne by the municipality or that the municipality has to compensate for the effects of these actions. It is not popular for a local authority to apply such strong and effective instruments and they are seldom used.

8.2 Finance, tax and subsidies

Certain subsidies specific to the property market are outlined here. Otherwise, the financial, tax and subsidy environment and information systems are as described in respect of the land market in Chapter 4.

In addition to the tax incentives described in Chapter 4.3, which are also valid for the property market, there are some specific tax regulations applicable to the existing building stock.

In 1977 tax policy changed because of the decreasing population in the inner cities and because of poor-quality or unoccupied building stock. The tax relief available for new buildings was extended to the existing building stock. This was particularly important for owner-occupiers. Tax reliefs for owner-occupied flats and houses since 1977 apply to purchases of second-hand dwellings in the same way as for new ones. This is the most important tax regulation affecting the property market and it encourages purchase rather than renting of a flat. Today, when new flats are scarce, the government is moving towards cutting back these tax concessions relating to older dwellings in order to create greater financial incentive to promote new construction.

Extension of residential buildings is promoted in order to reduce scarcity in the housing sector in the following ways:

○ a landlord is subsidized by being allowed annual tax deductions of 20% of the cost of additional flats offered in existing buildings. The costs of the extension are limited to DM60,000 per new flat so the maximum annual deduction is DM12,000 over a period of five years (§7c EStG);

○ tax concessions for owner-occupied dwellings may also be taken up for building extensions.

Costs of modernization and maintenance may be subsidized in the following ways:

○ the owner-occupier of an office or factory building may deduct all these costs from his taxable income as costs incurred in keeping the company running;

○ a landlord, regardless of the use of a building (office, industrial or residential), may deduct the costs of modernization and maintenance from his taxable income. The rent law prescribes that modernization of flats may result in a rent increase of not more than 11% of the cost of modernization per year.

However, the owner-occupier of a flat or single house may not deduct these costs from his taxable income because a person's own flat is considered to be a consumer item.

Special tax reductions of 10% over 10 years are also possible for the construction costs of energy-saving and ecologically desirable heating systems (§82a EStDV).

The owners of buildings classified as listed buildings or monuments are eligible for tax deductions. The costs of conservation may be deducted by 10% annually over 10 years, for landlords as well as for owner-occupiers (§7i and §10f EStG). Modernization and maintenance of buildings in redevelopment areas is promoted in the same way as listed buildings (§7h and §10f EStG).

All these allowable deductions are important for the development of the existing building stock. Without them, investment in the building stock would not often be profitable. Other subsidies affecting the property market are as described in Chapter 4.3.

CHAPTER 9
The property-market process

9.1 Price-setting

There are, however, changes in prices and values within the property market as compared to the land market. The prices on the property market depend less on public action, and more on market situations. Economic activity, and demand and supply of property, are the main factors causing changes in property values.

Public-sector activities also affect property values. For example, a new public transport link to a residential area on the edge of a town will normally increase the price of the properties there. Public-sector provisions for improving conditions in a city district or in a block are also important. These might be a new function for a district or neighbouring district within the structure of the town, provisions to calm traffic and to improve the housing environment of a block or district (Wohnumfeldverbesserung), or the redevelopment and renewal of residential and industrial areas (Sanierung). These publicly supported provisions often result in increasing property values or at least their stabilization. The change in values caused by public actions such as the listing of buildings or prohibition of conversion into flats or apartments (Zweckentfremdungsverbot) is variable.

The methods of valuation are the same as those described in Chapter 5.2. In the property market, however, the comparative method is not used very often because data on comparative objects of valuation is not readily available. The return method and the cost method are most commonly used.

9.2 The actors and their behaviour

In the property market the same groups of actors are involved as in the land market. Their aims, motivation and behaviour are the same as in the land market and as described in Chapter 5.2.

The public actors are less important in the property market. Here they do not have the function of controlling supply, and they seldom participate. Their influence is more indirect, for example, through special activities and legal instruments as described above or through taxes and subsidies.

In the housing sector, rent legislation sets restrictions on rent levels and increases. The privately organized tenants' associations (*Mietervereine*) often publish lists with typical rents per town (*Mietspiegel*). In the industrial sector there is no rent control.

The suppliers of second-hand properties are mainly private owners. Most sell their property only if the household wishes to move. Some of them sell in order to buy a larger or more comfortable dwelling. A special group, often supported by property professionals, are owners who intend to convert rented flats into apartments. Often they begin as buyers in the property market and then become suppliers of apartments (see the Köln case-study, Ch. 11.3).

Special groups of actors are involved in redevelopment projects. Within the residential sector there are public or semi-public institutions acting in trust in renewal areas. Most of the Länder and the larger towns have founded such corporations. The large housing associations are also engaged in this field: for example, the Landesentwicklungsgesellschaften (LEG), the Deutsche Stadtentwicklungsgesellschaft (DSK), the Stadtentwicklung Südwest (StEG) Stuttgart, and the Stadterneuerungs- und Stadtentwicklungsgesellschaft (STEG) Hamburg. These actors are usually not intermediate owners of the land, except in the case of some prime plots.

The redevelopment of industrial locations is an important task, especially in the older industrialized regions. Here the institutions referred to above are also active, the Grundstücksfonds Ruhr within the LEG of Nordrhein–Westfalen being a good example. In this field, private companies, such as the large construction companies, developers, banks or new societies founded by these actors as shareholders, are also involved (see the Düsseldorf case study, Ch. 11.1). In such cases redevelopment takes place in public–private partnerships. These actors are usually intermediate owners of the redevelopment area.

Those demanding property are the same as described in Chapter 5.2. In the property market the largest group of demanders are tenants, not only in the housing sector but also in the office and the high-tech industrial sectors.

The description of the property professionals in Chapter 5.2 are also valid for the property market. Nevertheless the developers, building contractors and construction companies are less important to the operation of the property market than to the land market, except within redevelopment projects. The estate agents and the banks are the most important professionals in the property market. Most estate agents in Germany are active in the property market, not the land market. Where conversion of rented flats into apartments is concerned, estate agents, architects and lawyers are of specific importance (see the Köln case-study, Ch. 11.3).

The housing associations are also important actors on the property market because of their large building stocks. They supply and manage thousands of rented flats for tenants with low incomes, playing the rôle of landlords.

CHAPTER 10

The outcome of
the urban property market

10.1 Demand and supply of property

In the past 10 years the demand for residential properties has responded to general conditions in the housing market. At the beginning of the 1980s the demand increased, while in 1985 and 1986 there was a steady collapse in demand because of falling incomes and oversupply, and prices declined. As in the land market, the situation has since changed and demand for properties is currently outstripping supply.

Demand takes place predominantly in the existing housing stock, with 80–90% of all purchases being existing previously occupied houses or apartments. It is expected that demand for used properties will continue to grow until 1995.

The demand for used properties is stimulated by a general desire for ownership and to invest capital in property. Because many households are not able to finance new houses, they have been urged to buy second-hand prop-

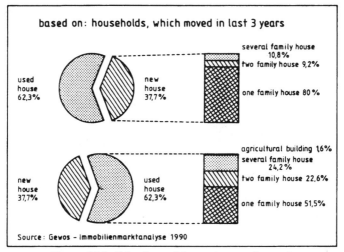

Figure 10.1 Behaviour of demanders on the market.

erties. Figure 10.1 explains the behaviour of consumers (based on households) and the ratio of purchases of used and new houses and apartments. In general, buyers are more flexible between new or used property than between different types of house. In addition, many households today are urged to buy properties because rents are relatively high and in some regions buying can be cheaper than renting.

The figures do, however, vary over time. Raschke (1991), for example, indicates that in 1985 the volume of turnover of used and new properties was equal, but by 1987 used-building turnover accounted for 67%, and in 1989 the ratio was 44% new buildings to 56% used.

The largest demand for houses, and also the largest supply, is concentrated in the conurbations, especially the surrounding countryside, and in regions with a high rate of immigration, for example in the south German conurbations of Rhein–Main, Stuttgart and München. Although these regions have a higher level of demand, for the past two years more northern regions such as Hamburg and Berlin have shown above-average rates. It can be predicted that the demand will switch more to the less-populated surrounding countryside, because in central urban areas and in the more highly populated rural areas prices will be high and supply insufficient.

10.2 Transactions in the property market

Each new building transaction today is matched with three used-building transactions. In 1990 the turnover of used properties (houses and apartments) was approximately DM180 billion and has increased continuously during the 1980s. A yearly turnover of DM200–250 billion is to be expected until the mid-1990s, particularly since the proportion of inherited properties is growing and the average size of households is declining (Preisinger 1991: 31). This trend will be supported by the introduction of the Single European Market in 1993.

In future the turnover of used properties will also grow because changes are likely in the duration of ownership. While the average duration today is approximately 28 years, it is predicted that it will become shorter because households of younger generations are more mobile and have less emotional ties to property (Preisinger 1991).

The number of transactions also differs between types of used property. In particular, the number of transactions for two-family houses has declined, while the growth of single-family homes and apartments has stabilized.

Generally, used properties are traded much more in large cities, while new buildings dominate in small towns and in the rural areas. Apartment transactions, especially in large cities, are becoming more important. In 22 selected German cities in 1989, the share of apartments among all other transactions was 56%, compared with only 52% in 1987. Nevertheless, the share has levelled since 1988 (see Fig. 6.2) (Hüttenrauch & Schaar 1990: 691).

Normally the market analysis for apartments is subdivided into purchases of new, used and converted apartments. Converted apartments are defined as former rented flats. Transactions involving used apartments in northern Germany declined from 67% in 1987 to 48% in 1989, while the share of new apartments levelled out at 12% and the conversion from rented flats into apartments increased from 27% to 40%. In southern German cities, by contrast, the share of new apartments was 48% in 1987 and it declined to 30% in 1989. By contrast, the used-apartments sector grew from 34% in 1987 to 42% in 1989, and converted apartments levelled out at 28–9% (Hüttenrauch & Schaar 1990). It is to be expected that the shares of new and converted apartments will increase slightly. For example, nowadays approximately 70,000 to 80,000 flats per annum are converted or demolished.

Table 10.1 Share of transactions of new and used buildings in 1989 compared to the size of a town (percentage).

Population	new property		used property	
below 20,000	55	32	45	25
20,000–100,000	49	39	51	38
100,000–500,000	46	22	54	25
above 500,000	36	7	64	12
		100		100

Source: Raschke, W.-D., In: Der Langfristige Kredit,8/1991:258.

10.3 Prices

Data about prices is given only for residential units, because no exact data for the commercial sector is available. For rents, see Chapter 1.4. As mentioned before, the prices for used properties increased at the beginning of the 1980s, whilst there was a decline in the mid-1980s. The changes were rather more pronounced in the apartment market (see Fig. 10.2). There have been no great price differences within regions, although house prices as well as apartment prices increased slightly more in the surrounding countryside than, for example, in the central urban areas.

The frequency of transactions in different price ranges differs for single-family houses and apartments. While one- and two-family houses sold in various German cities for average prices of up to DM300,000, apartments sold at average prices of up to DM200,000.

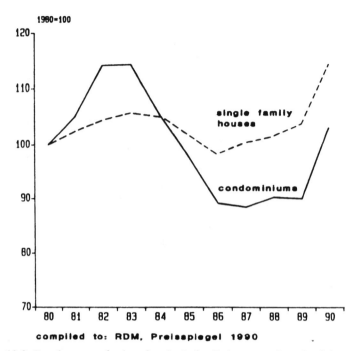

Figure 10.2 Development of prices for single family houses and condominiums 1980-90.

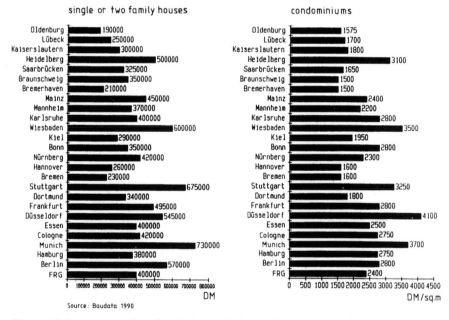

Figure 10.3 Average prices for single family houses (total price) and condominiums (in DM/m²) in several German cities in 1990.

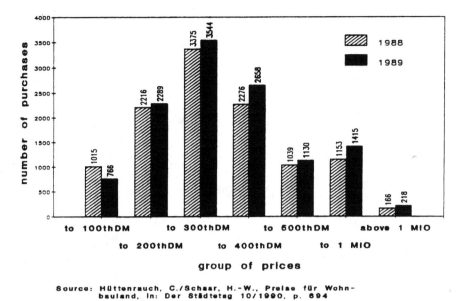

Figure 10.4 Frequency of transactions for one and two family houses differentiated between groups of prices in 1988 and 1989.

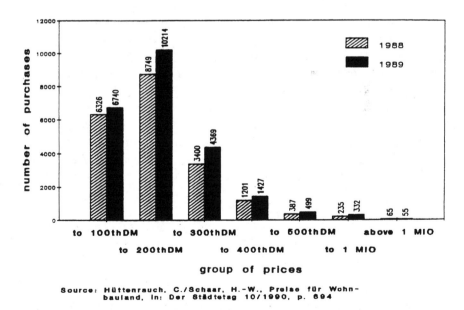

Figure 10.5 Frequency of transactions for condominiums differentiated between groups of prices in 1988 and 1989.

CHAPTER 11

Case studies of the property market

11.1 Düsseldorf

Introduction

Like many other cities Düsseldorf is still affected by the legacy of generations of industry. Many manufacturing companies have closed down production facilities in the inner-city area. Once settled in the outskirts, as a result of suburbanization, they have been integrated into the surrounding urban environment. Lacking expansion potential or the flexibility for structural change within their branches, they have been inevitably forced to close down, creating a problem of derelict industrial land. Despite being attractive locations these areas have frequently remained unused for a long time. Vacated buildings are rarely used for other purposes because they are not up to the standard required by modern organizations. Moreover, the soil is often very highly contaminated and must be cleared. Since the shortage of industrial land has become more acute, derelict industrial plots have come to be considered a major source. This, however, does not take into account the question of the image of a company, and the possible negative effect of locating on former derelict land rather than greenfield sites.

These particular sites are often excellent development sites because they are close to the inner-city areas and are well supplied with infrastructure. Regionally important infrastructure such as motorways and railway connections exist in many cases. The Hansa Park case study is a representative example of a new generation of industrial property development.

Besides the location, industrial park projects depend on planning permission procedures not causing delays. Therefore private investors call for expeditious administration of the law and unbureaucratic consideration of projects.

One legal instrument capable of accelerating the siting of an industrial park in an inner-area location is §34 BauGB (admissibility of projects within inner-area locations) in order to grant permission, in order to avoid a lengthy Bebauungsplan procedure.

The Hansa Park project was a prototype of this procedure, because until the early 1980s industrial parks were normally built on greenfield locations in the outer conurbation area and not in inner-area locations.

The geographical and economic situation　As the capital of the most populous Land in Germany, Nordrhein–Westfalen with 17 million inhabitants, the city of Düsseldorf enjoys a special status. It has about 570,000 inhabitants and is surrounded by highly populated conurbations. To the south and west there is the so-called conurbation of Rheinschiene, which includes the area of Köln and Bonn and the city of Leverkusen, and to the west there are also the cities of the Lower Rhein: Mönchengladbach and Krefeld. In the north is the conurbation of Ruhrgebiet, which includes the cities of Duisburg, Mühlheim, Essen, Bochum and Dortmund. To the east of the Greater Düsseldorf area are the towns of Wuppertal, Solingen and Remscheid.

This geographical position also gives Düsseldorf a special status in Europe. The whole agglomeration of the Ruhrgebiet and the Rheinschiene is the largest in Europe. Within a radius of only 50–60 km about 10 million people live and work. In other words, Düsseldorf is located in the centre of the northwest European settlement area, which is only a short distance from the Benelux countries. It makes up one of the largest self-contained markets and has high spending powers. One in seven inhabitants of the EC live within a radius of about 200 km. (Wirtschaftsförderung Düsseldorf 1990). The city of Düsseldorf belongs to No. D 51 of the European Region System and is termed Region Düsseldorf.

The city of Düsseldorf is the *Schreibtisch des Ruhrgebietes* (the "desk of the Ruhrgebiet"). This is because it used to be renowned as the administrative and service centre for the production industries (such as coal-mining and steel making) in the Ruhrgebiet. Now the city has developed into an international trade and service centre and there has been a continuous shift in the labour market from production to the service sector. The market share of the service sector has increased to more than 70% (compare with the Frankfurt case study in Ch. 11.2), while jobs in industry, crafts and the building trade have declined from more than 50% in 1950 to less than 30% today (Neisser 1989).

The industrial location of Düsseldorf　Above all, it is Düsseldorf's position in this extraordinary area that makes the location so attractive for domestic and foreign companies. All the neighbouring municipalities realize that they can profit by their favourable position adjacent to Düsseldorf, which is one of the top business locations in the EC. For this reason the neighbouring municipalities co-operate with the city of Düsseldorf and in the course of the publicity campaign "Düsseldorf & Partners" the whole region Düsseldorf/

Middle Lower Rhein has been promoted as a business location with a great future (Landeshauptstadt Düsseldorf 1989).

Apart from 90,000 German companies, more than 4,000 foreign companies are sited in the economic region of Düsseldorf. Traditional trading partners, such as the USA, the Netherlands, the UK and France are represented and about 320 Japanese companies with more then 7,000 workers have settled here. So Düsseldorf competes with London as the continental European control centre for Japanese economic activities in Europe. Other East Asian growth nations such as Korea and Taiwan are also represented.

Even with respect to east–west trading connections, which will become more important because of changes in Eastern Europe, Düsseldorf plays a leading part. This is illustrated by the fact that Russia intends to establish its first trading centre in western Europe in Düsseldorf (Wirtschaftsförderung Düsseldorf 1990).

In its promotion as an international business centre, Düsseldorf is working on establishing an "international infrastructure". The area around Düsseldorf has excellent transport connections, and all European business centres are readily accessible. The Rhein–Ruhr Flughafen is the second largest commercial airport in Germany, with connections to 150 cities worldwide, and is also an attractive location for different industrial uses. The Düsseldorf Fair Centre, another important feature, offers a programme of 35 international trade fairs (Wirtschaftsförderung Düsseldorf 1990).

Industrial land market and policies The special position of the economic region of Düsseldorf and the high level of demand for industrial land leads to the question of whether the region has an adequate land supply.

According to a statement of the Wirtschaftsförderungsamt (economic assistance committee), the land market of Düsseldorf is overheated because of high demand. There is virtually no stock of industrial land immediately available: at the time of writing there was only an 8.9 ha plot of vacant land. The new Flächennutzungsplan has just been submitted for approval to the president of the Regierungsbezirk. This allocates 210 ha of industrial land for release, with contaminated land as potential building land not accounted for. By contrast, the predicted demand is of the order of at least 280 ha for the period up to 2000. But only 85 ha of industrial land can be developed by then, about two years' supply at current demand levels.

In the office market, the centre of Düsseldorf has an excess of demand over supply that will last for many years. The outskirts of the city centre and outlying districts are therefore becoming obvious alternative areas for office development. Map 16 illustrates industrial locations for business parks.

Even in the industrial land market, there is a tendency to move to the outskirts in the economic region of Düsseldorf. For this reason, the economic

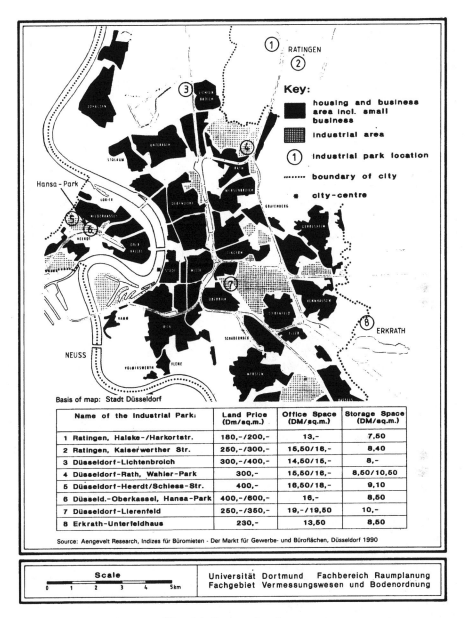

Map 16 Business locations.

assistance committee has sought to dissuade companies from moving to other regions and to encourage them to remain in the Düsseldorf area.

New policies promote the Düsseldorf/Middle-Lower Rhein region in the "Düsseldorf & Partners" publicity campaign. The neighbouring municipalities provide the industrial land that Düsseldorf lacks. So a total supply is to be created that is to be regarded as a unit, an arrangement seen to be advantageous to all concerned (*Düsseldorf Magazin* 1989: 44–5).

Subsidies or grants are hardly necessary for the Düsseldorf region. First, area does not belong to a region of Germany or Europe entitled to structural aid subsidies. In addition, subsidies or incentives for development from the municipality are not necessary because of high levels of demand.

This also applies to the supply of industrial land for development on the sites of former industrial plants. The municipality of Düsseldorf does not need to act as an intermediate owner carrying out decontamination of soil on its own account in order to create a supply of high-quality industrial land. The Düsseldorf industrial land market enjoys such strong attraction and demand that high purchase prices for sites that still need to be cleared can be obtained. This means that the cost of cleaning the contaminated soil is borne by the original owner in the form of a reduced sale price (*Düsseldorf Magazin* 1989: 17). This situation is totally different from the conditions confronting the municipality of Dortmund (see Ch. 7.3).

The Hansa Park project

Description   The site of the case study is an industrial park on a former manufacturing plant of the Thyssen steel company. The Hansa Park project is located in the municipal district of Oberkassel, less than 6 km from the centre of Düsseldorf (see Map 16) and is a representative example of industrial property development and investment by a private developer. It has provided the pattern for many similar developments in other growth areas, which until recently have developed as an almost independent market involving various new actors within the planning process.

In late September 1982 the Trammel Crow Private Limited Company, from which the Calliston Private Limited Company emerged soon after, bought 84,000 m^2 of the site. The original owner was bound by contract to bear the cost of demolition and removal of contaminated soil, which involved clearing 30 cm of the surface soil. At that time, the land price was DM100–150 per m^2.

The new buildings, which had a net internal area of 58,000 m^2, were completed in eight phases of construction. Office space takes up about 53% of this, and warehouse space about 47% (Map 17).

The method of construction was not typical. A unique method of construction with prefabricated elements (flexible buildings) enabled the Calliston Management Company to adapt the office and warehouse space to the particular requirements of tenants. Furthermore, variations could be made for tenants' changing floor-space requirements, or in respect to a change of tenants. Calliston was able to react to the requirements of the new tenant, securing

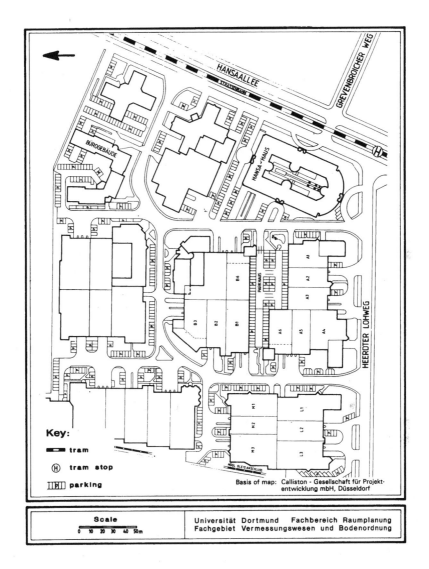

Map 17 General location.

the long-term profitability of the buildings and having the option of increasing rents.

Four architects collaborated in designing exclusive facades that incorporated clinker, natural stone and glass elements, thus creating an attractive effect. The varied landscape design was another element satisfying tenants' demands for a high-quality image and a pleasant environment, and at the same time increased the value of the site. The proportion of green space is around 25%. Conflicts with the surrounding residential area are negligible.

Altogether, 33 high-tech companies specializing in electrical engineering, environmental engineering and information technology are located in the Hansa Park. Many Japanese companies have set up their German head office there. The 1,500 jobs created in the Hansa Park have at least mathematically compensated for the job losses resulting from the closure of the Thyssen steelworks (Kunzmann et al. 1986).

Originally the buildings had monthly rentals of DM12.50–14.00 per m². But these have now risen to DM20–22. The strategy of the management company seems to have worked very well. With lease contracts of just five years, the project is planned in the expectation of multiple lettings of the buildings. The profits are not generated speculatively within a short period, but are expected to be created in the longer term as rents rise. The success of this project is also apparent from the fact that Calliston has already developed a second industrial park (Wahler Park) in the Düsseldorf region, based on the concept and experience of the Hansa Park.

Stages of development At the beginning of the 1950s the Thyssen Cast Steel Limited Company had built a steel plant on the Hansa Park site. In 1980 the company closed this site for structural and political reasons. For the following two years the area lay derelict (Fig. 11.1). Buyers or users for the large halls, shops and administration buildings could not be found. The former industrial buildings became unused or were only used temporarily (for example, as a theatre-workshop and as a warehouse).

Only two years later, Trammel Crow GmbH became a prospective buyer of this derelict area after an estate agency had drawn its attention to the site.

Parallel to the purchase negotiations, investigation of the extent of contaminated soil was already in progress. It was not carried out to today's standards, because the problem was not as well understood in 1982 as it is now, and secondly because the proposed use of the area had not been classified as appropriate for potential contaminated land.

In the plan for the Hansa Park industrial park, Trammel Crow applied to the municipality of Düsseldorf for a building permit soon after purchasing the area in early September 1982. In order to prevent a delay because of bureaucratic procedure, and also to accelerate the granting of building permission, Trammel Crow insisted on calling all the private and public actors in the planning process to a round-table conference to discuss the Bauvoranfrage (preliminary application) for a building permit. Within four weeks, Trammel Crow had received the municipality's comments and planning permission according to §34 BauGB. Usually this process takes at least 6–10 weeks. Thus, obtaining permission under §34 BauGB instead of a building-plan procedure accelerated the project by between 8 and 10 months.

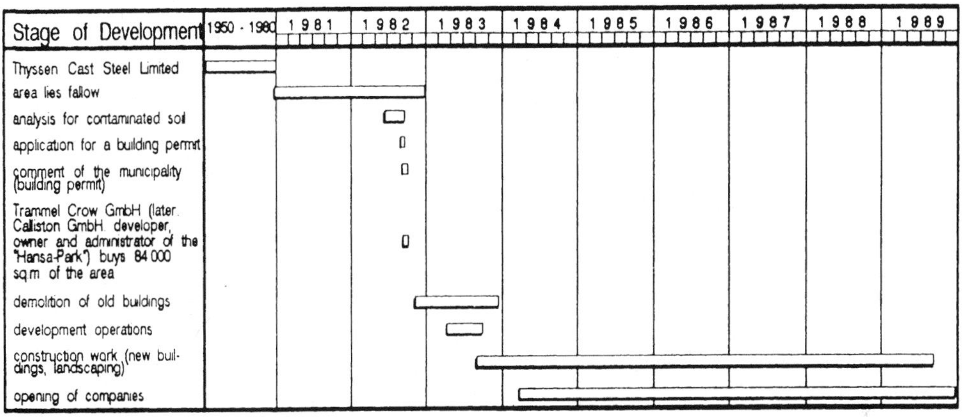

Stage of Development	1950 - 1980	1981	1982	1983	1984	1985	1986	1987	1988	1989
Thyssen Cast Steel Limited	▭									
area lies fallow		▭								
analysis for contaminated soil			▭							
application for a building permit			▯							
comment of the municipality (building permit)			▯							
Trammel Crow GmbH (later Calliston GmbH developer, owner and administrator of the "Hansa-Park") buys 84 000 sq.m of the area			▯							
demolition of old buildings				▭						
development operations				▭						
construction work (new buildings, landscaping)				▭						
opening of companies				▭						

Figure 11.1 Timetable, Hansapark.

Under the centralized control of Calliston, the project development company, development operations for the Hansa Park site were started in spring 1983. Calliston owns the whole of Hansa Park, including the buildings and the streets. This gives the company greater flexibility with regard to the use of the site. Parallel to these development operations was the demolition of the former Thyssen buildings, which were still partly occupied by interim users.

Development of the first offices and warehouse buildings began on 2 August 1983, and in April 1984 Pioneer was the first firm to open its central office in the new Hansa Park.

The development process: actors and their instruments

The property professionals: the developer and estate agents Private companies have been decisive parties in the smooth running of the Hansa Park project, with Calliston, the project developer and exclusive owner, holding the key position.

The transformation of the former industrial location to an attractive industrial park for high-tech firms has been exclusively carried out by the developers. Subcontractors have been called in only for some demolition and construction work.

The plan in this case was prepared exclusively by the developer, which had already gained much more experience of such projects than the planning authority, and had carried out well targeted market research. Because it was the only investor and the exclusive owner of the Hansa Park, it was important for Calliston to develop a project that would secure long-term profitability of the buildings. This is an important requirement of the municipality when giving approval to an industrial park according to §34 BauGB, because the quality of the area is secured in this way.

The project developer had to press for planning certainty and an acceleration of administrative procedure to enable lease negotiations with potential tenants to begin. For this reason an estate agency was involved, which frequently acted as a negotiator in the course of the project and set up the round-table conference referred to above. This arrangement, which brought about considerable acceleration of the procedure, is as a rule unusual in the planning process.

The estate agency first became involved in the Hansa Park project when acquisition of the derelict industrial land was under consideration, and during this stage the municipality was of little importance. Furthermore the estate agency acted as a mediator between municipality and developer during the approval procedure. The estate agency even took charge of the acquisition of tenants. Contacts were also arranged by the economic assistance board.

One tenant wished to withdraw from a lease contract before legal completion, because the rented building did not suit his increased floor space requirements as a growing company. Calliston therefore negotiated the transfer of this particular company to another industrial park that it had developed and owned, in order to provide it with a suitable building.

The rôle of the municipality The city of Düsseldorf has not been the main actor in this project, but has still made an important contribution to the acceleration of the planning permission procedure. As noted, the legal instrument accelerating the planning of the Hansa Park project and avoiding a Bebauungsplan procedure was the grant of permission according to §34 BauGB.

"There is nothing wrong with a planning permission according to §34 BauGB, if the project fits in with the particular character of the closer environs with respect to the type and size of the physical use, the type of coverage, and the plot of land to be built on", commented the representatives of the municipal economic assistance board. "The developer presents a complete plan of an industrial park, which would be a large-scale and long-term duty for the municipal planning department. This can be the reason for a delay of the project." (authors' translation).

These municipal representatives also observed that a building-plan procedure might even be a hindrance to the interests of the public economic development board. In the case of the Hansa Park the regular dialogue between the developer and the municipality facilitated the extraordinarily fast planning of the project. The municipality of Düsseldorf was not involved in the land purchase, site-clearing or redevelopment of the area, and the economic assistance board refers prospective firms to the manager of the park.

Local measures for the purpose of revaluing the surrounding area obviously had little influence on the success of the Hansa Park idea, in the opinion of the park's manager. Other important locational requirements – in particular good traffic links, services and infrastructure in the environs – are no substitute for that (Kunzmann 1986).

Because of the attractiveness of the region there is substantial private-sector development potential in the Düsseldorf industrial land market, and so the municipality defines its rôle in the process of industrial development rather cautiously: "In fact, local economic development policy can only support private initiatives. Not by means of grants, but by assistance in form of planning, and by help in getting over bureaucratic statutory hurdles. Strictly speaking it is very important that the aims are clear. Only then can economic development be pushed ahead with municipal help. In other words they can only passively indicate to prospective companies that their intentions are corresponding with the local aims. In this case the municipality can adopt supporting measures to speed up the desired developments" (*Düsseldorf Magazin* 4/1989: 17).

The tenants Hansa Park provides premises for 33 high-tech companies. Most are internationally respected Japanese firms that have opened their German or even their European headquarters at this location. For this reason high-quality buildings within an attractive environment were demanded in order to present a good image to their clients and customers and satisfy the requirements of staff.

All these companies rented their buildings but Hansa Park makes clear that renting an industrial building does not preclude flexibility for the tenants.

Another reason why foreign companies predominate in the Hansa Park is that they are inexperienced in the German industrial property market and therefore welcome as much help as possible. Everything is arranged by the developer through the manager of the Hansa Park. The tenants just have to negotiate with Calliston about the lease contract. Special requirements expected of the tenant are embodied in this contract.

11.2 Frankfurt

The general setting

Frankfurt in the west German economy Frankfurt plays an important rôle in the western German economy since it is a major location for company headquarters, especially those in the financial sector. There is more decision-making power concentrated in Frankfurt than the number of headquarters might suggest. Moreover, it is in Frankfurt that the key data for the German economy are set. The Bundesbank, which is largely independent of government in controlling interest rates and money supply and deciding about interventions on the foreign currency market, has its headquarters there.

More than 400 banks are represented in Frankfurt, more than 60% of which are foreign. In 1988, 79 out of the 149 German banks had their headquarters there. The Frankfurt stock exchange is the most important in Germany, with more than two-thirds of Germany's securities transactions taking place there (Heinz 1990: 123). Frankfurt's trade fairs are of international significance. In 1989 there were 32,482 exhibitors at 25 events, 16,512 of which were from abroad, and about 2.6 million visitors, 10.6% from abroad (Stadt Frankfurt 1990a: 81).

Frankfurt and the Rhein–Main agglomeration (Map 18) is a leading international distribution centre because of excellent national and international transport and communication systems provided by the international airport, fast motorway and rail links. The city and its region is also a centre of industrial production, providing 98,000 jobs in industry, of which 40,000 are in the chemical industry and about 27,000 in the electrical industry (Stadt Frankfurt 1990a: 52).

Thus, the key function of Frankfurt and the Rhein–Main agglomeration may be described as a switchboard for the international exchange of information, capital and goods between Germany, Europe and the world.

While sustaining and expanding its international position, Frankfurt no longer competes with other German cities, but rather with other European capitals for the top rank in Europe as a banking and financial centre. This ambition is underlined by its efforts to become the location for the proposed European central bank.

Economic and social structure of Frankfurt and the Rhein–Main agglomeration The Frankfurt and Rhein–Main agglomeration is part of the state of Hessen. Hessen's southern subdistrict (Regierungsbezirk Darmstadt) broadly covers the Rhein–Main agglomeration and represents the EC region of Darmstadt (No. D61). The city of Frankfurt and some surrounding districts form the Umlandverband Frankfurt, which is the competent authority for land-use planning, waste disposal, water supply and other functions (Map 18).

Compared with other European cities Frankfurt is quite small. In September 1990, 632,535 persons lived in an area of 248.36 km^2 (Stadt Frankfurt 1990b). But the city of Frankfurt proper covers what can be regarded as only the inner city of a larger conurbation. Within this area there is an internal division of functions that results in strong interdependence between the diff-

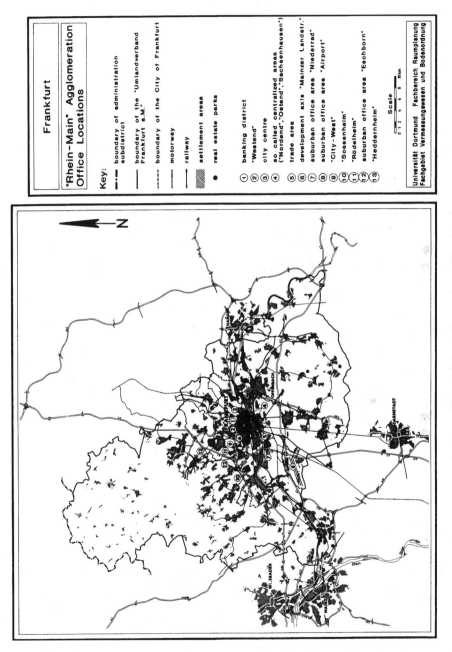

Map 18 Rhein-Main agglomeration: office locations.

erent parts. The Umlandverband Frankfurt has a population of 1,551,211 in an area of 1,427 km². The Regierungsbezirk Darmstadt has a population of 3,491,419 in an area of 21,114 km² (Stadt Frankfurt 1990a: 191).

In Frankfurt there are 558,852 jobs (21.9% of all the jobs in Hessen) and 76.7% of these are in the service sector, the highest rate in Germany (Stadt Frankfurt 1990a: 52). Of the 907,782 jobs in the Umlandverband Frankfurt, 71% are in the service sector (Umlandverband Frankfurt 1989: 7). To fill the jobs 283,451 commuters come to Frankfurt each day. Since the 1970 census the number of commuters has increased by 199,176 or 42.3% (Stadt Frankfurt 1990a: 42). The rate of unemployment is relatively low: in September 1990 it was 4.6%, compared with 5.3% in Hessen and 6.6% overall in Germany (Stadt Frankfurt 1990a: 85).

Although Frankfurt seems to be a very prosperous city, there has been a rise in the number of people on income support. Between 1980 and 1988 the figure increased from 5.8% to 8.7% (Institut für Raumplanung 1989: 7; Stadt Frankfurt 1990a: 118). There is also a shortage of affordable housing. A high demand for inner-city flats has led to steep rent increases in inner-city residential areas. Monthly free-market rents spiralled up from about DM7.00 per m² in 1980 to about DM20.00 per m² per month today. Since the mid-1980s, the social housing programme has been reduced from about 1,700 flats per year to 300. Meanwhile there is a housing problem for all types of demanders and companies are finding it more and more difficult to attract highly skilled employees.

Strategic policies for future development of the city and region Frankfurt's city council aims to support economic growth of the city and to extend its international importance, but without neglecting social and ecological problems caused by the expansion of the service sector. It tries to provide space for new office development in existing commercial areas, especially in the traditional banking district. Simultaneously, it seeks to protect – from gentrification, from conversion of flats into apartments and from commuter traffic – residential areas adjacent to areas where offices will be developed. In areas for office development, the city council hopes to encourage the provision of an office and housing mix. It has decided to allow for dense office development along the Theodor-Heuss Allee adjacent to the trade-fair area and to permit additional office towers in the traditional banking district in order to reduce the pressure there for office development.

The state of Hessen and the regional planning authority, the Regierungs-Präsident Darmstadt, have regional development policies that seek to support the concentration of the service sector in the centre of the Rhein–Main agglomeration and to allow the expected relocation of industry from out of the central area. They wish to encourage the relocation of this industry in designated growth areas in the state's periphery. So far these policies have failed, because industrial firms have preferred to relocate in new locations within a small radius around Frankfurt.

The 1985 structure plan of the Umlandverband Frankfurt had to comply with the policies of the regional plan (Regionaler Raumordnungsplan). Thus its provision of employment areas is designed to accommodate no more than a constant number of jobs but to allow for internal mobility of firms. The eastern part of the area covered is particularly pinpointed for population and employment growth. It is the policy of the planning authority to sustain the existing division of labour between centre and hinterland (Umlandverband Frankfurt 1988).

The Frankfurt office market

The information presented in this section is based on interviews with officials, yearly reports published by real-estate consultants and from a study by Hennings et al. (1990) for the city of Offenbach.

Functional and spatial segments The Frankfurt office market is differentiated into functional and spatial segments in terms of demand and supply:

○ Several business parks sited in the region in locations that provide good access to the motorway network (see the Düsseldorf case study, Ch. 11.1). The provision of office space in business parks on a larger scale is the latest development. Occupiers of offices are companies involved in distribution. Other sources of demand for these locations include marketing, service and research and development units of the industry, depending on the quality of the office development.

○ The international airport is a major source of demand for office accommodation. It is a major source of employment, with more than 40,000 employees on the site. The service functions will be expanded as part of a DM7.4 billion development. Additional office floorspace for up to 15,000 employees is in the pipeline, to be developed by the year 2000.

○ Financial headquarters functions, such as decision-making, and foreign currency and securities trading, tend to be housed in high-density locations in the "banking district", which is close to the stock exchange on the western edge of the medieval city core. Financial-sector firms seek the most central location available.

Around these business cores there is a spectrum of service-related firms that try to get as close to their customers as they can afford.

Within Frankfurt there are central and peripheral locations (Map 18). Central locations include the banking district, the Westend and the Inner City. Other areas about to be "centralized" include Mainzer Landstraße, Bahnhofsviertel, Bockenheim-Süd, locations adjacent to the trade fair area,

217

Sachsenhausen and Nordend. In these locations, except the banking district, beside banks and insurance companies one can find offices of architects, solicitors, consultants, branches of foreign companies, marketing advisers and personal services.

In the more peripheral locations, such as Niederrad, Eschborn, Heddernheim and Rödelheim, there is a mixture of industrial and trade company's headquarters and service units, such as for high-technology products. Originally, most of the office buildings in Niederrad and Eschborn were owner-occupied. There are some new developments for the general office market. The peripheral locations of Frankfurt compete with other employment areas, especially business parks in the region for those demanders that rely on a good transport infrastructure, especially the airport. Other users are suboffices of banks or insurance companies. There is also a tendency towards the relocation of companies from more central locations because of rising rents or because their present accommodation no longer meets the requirements of modern office technology (Hennings et al. 1990: 133; Jones Lang Wootton 1990a: 7).

Demand Those seeking office property include those who want to buy, either for own use or for capital investment, and those who want to rent. Some acquire land for new office development, others are looking for modern office buildings, office buildings for refurbishment or office developments under construction.

The present demand for rented offices can be illustrated by the yearly take-up (Jones Lang Wootton 1990a: 5).

Year	1984	1985	1986	1987	1988	1989
Take-up (,000 m²)	110	143	260	200	200	270

Hennings et al. estimate the total demand between 1978 and 1987 (owner-occupied and rented) for offices in Frankfurt at approximately 2,400,000 m². For the period between 1987 and 2000 it is estimated that demand will be for between 2,600,000 and 3,250,000 m² (Hennings et al. 1990a: 159).

The investment market in Frankfurt has experienced several changes during the past decade. The most significant event was the growing involvement of foreign investors in the office market. Private investors who had previously dominated the market vanished from it in prime locations. Jones Lang Wootton estimate that the total property-investment assets of foreign investors in western Germany were about DM6.5 billion in 1988 and about DM11 billion in 1989 (Jones Lang Wootton 1989, 1990b).

Because land prices increase faster than office rents, office yields are continuously decreasing. In 1985 they were at 5.5% and in 1989 they were down to 3.5% (Gutachterausschuß für Grundstückswerte 1990: 78). Regulations on the activities of German insurance companies and open-ended real-estate funds have resulted in these investors being no longer able to compete with foreign investors in prime locations. They now have to prove a good return from rents, whereas foreign investors rely on the increase in property values to make their profits. The German insurance companies and real-estate funds have therefore diverted a major part of their investment sums to secondary locations.

Supply According to Hennings et al. (1990a) there is an inherent tendency towards an oversupply of offices. During boom phases, investors tend to develop offices on a large scale because of high rents and the expectation of further increases. However, there is a time lag of several years between the decision about the investment and the completion of a building.

In cases of oversupply and increasing vacancy rates, each supplier tries to solve his individual problem by reducing rents. But the level of prices has little effect on the quantity of demand for offices. Demand on the office market depends largely on the economic situation of final users. Therefore, the decreasing rents influence not the total vacancy rate but only the share available to investors in the vacant office stock (Hennings et al. 1990a: 147).

The vacancy problem is exacerbated when oversupply meets with a recession that results in decreasing demand. This situation occurred in the Frankfurt office market in the mid-1970s, when a boom of office development ran into the 1974 recession. Many office developments were begun before 1974 and were brought onto the market in the recession years. The result was an average vacancy rate in Frankfurt of more than 10% in 1976 and 1977. From 1981 onwards the vacancy rate was reduced to less than 4% by demand recovery (Jones Lang Wootton 1990a). New office development on a large scale began from 1983 onwards.

Hennings et al. (1990) also point out that local authorities have a decisive function in controlling the effects of the development cycle that market forces generate. By restrictive planning policies they could prevent vacancies. One of the experts interviewed took the same position. He stated that, for those who are already in the office market, a restrictive planning policy would help to sustain a shortage that would guarantee stable profits, while an oversupply would be hindered.

The total stock of office floorspace in Frankfurt is between 7–8 million m^2. The total floorspace of office units that are not integral parts of industrial plants or warehouses is estimated at 4 million m^2(Aengevelt 1990b). The annual additional supply of offices has varied over the past decade.

Table 11.1 Additional supply of office space (1000 sq m).

Year	1980	1981	1982	1983	1984	1985	1986	1987	1988	1989
Additional supply	143	157	95	90	110	120	260	200	200	210

Source: Aengevelt 1990.

Results of the market process The present imbalance between demand and supply has led to increasing rents, primarily in central locations and, more recently, in peripheral locations. Jones Lang Wootton give the figures shown in Table 11.2 for rental bands in different locations (Jones Lang Wootton 1990a: 11).

The "Messeturm", which lies within the so-called centralized area, is referred to as an exceptional case. Here the monthly rents of the latest contracts are DM70 per m^2.

The Gutachterausschuß für Grundstückswerte (valuation committee), reports average prices for office building land of DM21,850 per m^2 for sites with a floor-space ratio (GFZ) of 3.0 in prime office locations in 1990 (see also Map 19). In 1980 the figure was DM3,100 per m^2. However, real-estate consultants point out that no reliable market prices can be given for prime locations such as the banking district because of the rarity of sales.

The calculation of prices for office buildings as capital investment is based on the yearly net return from rents multiplied by a specified figure. The usual multipliers for different locations are (Deutsche Immobilienpartner 1990: 12):

Rhein–Main agglomeration   12–15
Periphery of Frankfurt   15
Central areas   17–20
Banking district   >20

Forecasts for future office-market development By the end of the 1980s, the question arose of how long the boom on the Frankfurt office market could be sustained. It was agreed that price rises would come to a halt in the early 1990s. It was expected that in those years the yearly growth of total office space would exceed the yearly demand, particularly in peripheral locations.

In 1989 the new city council issued statements criticizing three high-rise office buildings under consideration at that time. Speculation about a shortage of office space because of political restrictions on new office development emerged.

Since then, uncertainty over the three office buildings has diminished. Two

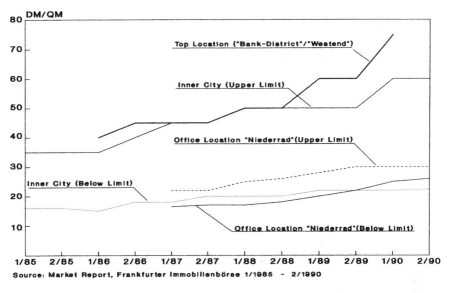

Source: Market Report, Frankfurter Immobilienbörse 1/1985 – 2/1990

Figure 11.2 Development of office rents (1985–90).

Table 11.2 Rental bands (DM/sq m /month) (cf. Map 18).

Year	1985	1986	1987	1988	1989	6/1990
Banking District	22–35	24–40	28–47	35–55	38–70	65–85
Westend	18–35	23–35	25–42	30–45	33–55	55–75
Central areas	14–30	17–32	18–35	20–38	22–40	35–55
Niederrad	16–22	18–22	18–23	19–24	22–32	26–35
Eschborn	15–18	16–18	16–20	22–25	22–26	22–29

Source: Jones Lang Wootton, 1990a:11.

of the three tower projects in question are under construction. In a framework plan for the banking district, an additional 191,000 m² of office floorspace is proposed, an increase of about 63%. Furthermore the city council has stated that it would permit further office development adjacent to the trade-fair area and along the Mainzer Landstraße.

Since 1990, when Germany's largest bank, Deutsche Bank, bought the Chase Manhattan Bank tower for DM180 million and the BfG tower under construction for an estimated DM1 billion, officials have been convinced that

Frankfurt will remain the financial centre of the German economy. While in the USA and the UK office markets are depressed, in Frankfurt there appears to be no crisis in sight, and an increasing number of foreign investors are entering the German market.

Estimates of future office supply vary:

○ Gaulke & van Mastrigt estimate an office supply of 1.6 million m² up to 1993 (Deutsche Immobilienpartner 1990: 11);

○ Jones Lang Wootton (1990a: 14) estimate the future completion of office developments until 1993 to be about 650,000 m²;

○ Aengevelt (1990) assumes that land reserves and local planning policies allow for office development at the scale of 2.5 million m² within the next 12–15 years, whereas up to 1993 they assume a supply of about 1,635,000 m²;

○ Hennings et al. (1990: 165) estimate that until 1992, 750,000 m² of office space will be completed. Additionally there are about 1 million m² under consideration.

Among experts it is agreed that future annual demand will exceed additional annual supply. Jones Lang Wootton (1990a: 6) estimate a surplus of demand until the end of 1991 at about 60,000 m². Thus, the vacancy rate will remain at a low level in coming years (Jones Lang Wootton 1990a: 5).

Some experts warn that there will be an oversupply from 1993 onwards. They estimate that a supply of between 300,000 and 400,000 m² will meet with a demand for only 200,000. They also warn that, if the German economy were to fall into a recession, vacancies would be very likely. Nevertheless, Hennings et al. expect no oversupply, especially in the long term, because restrictive land policies will prevent it.

There is a considerable difference in the expert forecasts concerning the future level of prices. With regard to the price level of London, which is more than three times that of Frankfurt, some state that a monthly rental of DM100 per m² will soon be surpassed. Others expect DM90 per m² per month to remain as a limit for the foreseeable future.

As a conclusion it may be stated that the relevant experts and actors on the office market appear to be confident that the boom will be sustained during the early 1990s. There are, however, some indications of a potential oversupply that might first affect peripheral locations. If restrictive planning policies should fail to slow down the pace of office development in the longer term, the market is most likely to produce more office space than demand is able to absorb. Another factor influencing the office market is the performance of the western German economy. Confidence and doubts about boom of the office market are based on different assumptions about that economy's future development.

Office development on the Mainzer Landstraße

This section illustrates the mutual influences of planning policies of the city council and the activities of investors. A closer examination is made of the development of the area along the Mainzer Landstraße, which is an inner-city main road (Map 18). Here many office developments and redevelopment projects are under construction or in the pipeline. The area in question extends from the edge of the traditional banking district to the city rail station of Galluswarte (Map 19).

City council policies In 1982 the city council commissioned a well known Frankfurt architect, Albert Speer, to undertake a study to provide guidelines for future urban development. Published as *City Leitplan* in 1984, it recommended a concentration of service-sector employment along three major development corridors. The study estimated the potential for new office development at between 360,000 and 550,000 m^2 until 1990 and 460,000 and 690,000 m^2 in the longer term. One development corridor was the Mainzer Landstraße. Speer suggested environmental improvement measures for the development corridors to attract investors. He also recommended architectural upgrading by high-rise buildings in certain locations. For the Mainzer Landstraße these plans were specified in a more detailed study in 1985 (Speerplan GmbH 1983).

By concentrating dense office development along inner-city main roads, the city council hoped to achieve three objectives: to provide sufficient office floorspace for the growth of the service sector, to keep office development pressure away from residential areas, and to overcome the historical development imbalance between the east and west of Frankfurt. Nevertheless, most recent developments indicate that the concentration of investment to the west of the city centre will continue.

By the time these strategic policies were developed, the following framework of legal plans existed. The Flächennutzungsplan of the Umlandverband Frankfurt designated the area largely for *Kerngebiet* (miscellaneous use). One exception was the site of a municipal car park adjacent to the Platz der Republik, which was designated as a site for public purposes. The *Bebauungsplan Westend* of 1978, which included the north side of the Mainzer Landstraße between Opernplatz and Platz der Republik, was intended to conserve the structure of the covered area in order to protect it for residential use. Therefore additional office development did not comply with the local plan. The remaining area along the Mainzer Landstraße is open to development according to §34 BauGB.

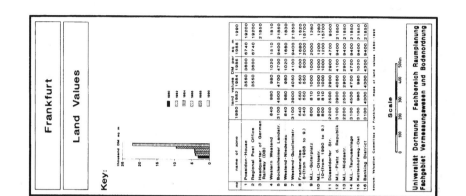

no	name of zone	land values DM per sq m					
		1980	1982	1984	1986	1988	1990
1	Poseidon-House			3550	3600	6740	19200
2	Regional Post Office			3550	3600	6740	19200
3	Headquarter of German Railway (DB)						21850
4	Western Westend	840	980	980	1020	1510	1810
5	Bockenheimer Landstr.	3100	4500	4700	4700	9400	21850
6	Westend-Niedenau	840	980	980	1020	1680	1830
7	Westend-Guiollettstr.	3100	3900	4050	4100	9400	21850
8	Frankenallee 8-(from 1988 to 9.)	540	550	650	600	1680	1520
		540	550	550	600	2000	15700
9	M.L.-Guterplatz	800	810	1000	1000	2000	1280
10	M.L.-Ostetr. 10-(from 1990 to 9.)	800	810	1000	1000	1200	15700
11	Dusseldorfer Str.	2200	2500	2900	2900	4700	9500
12	M.L.-Platz d. Republik	3100	3900	4050	4100	9400	21850
13	M.L.-Niddastr.	2200	2500	2900	2900	4700	21850
14	M.L.-Taunusanlage	3100	4500	4700	4700	9400	21850
15	Kettenhofweg-Ost	3100	980	1020	9400	9400	21850
16	Banking District	3100	4300	4300	4300	9400	21850

source: Valuation Committee of Frankfurt - maps of land values 1980-1990

Scale
0 100 200 300 400 500m

Universität Dortmund   Fachbereich Raumplanung
Fachgebiet Vermessungswesen und Bodenordnung

Basis of Map: Stadtvermessungsamt Frankfurt am Main

Map 19 Land values.

Development patterns of the Mainzer Landstraße Because of different historical development and locational qualities, the Mainzer Landstraße can be divided into the sections Opernplatz–Platz der Republik (referred to as Section 1) and Platz der Republik–Galluswarte (referred to as Section 2) (Map 19 and 20).

The real-estate consultants Müller International refer to Section 1 as part of the banking district. Jones Lang Wootton distinguish between the eastern part of the south side, which they regard as part of the banking district, the western part of the south side, which they regard as part of what they call "Centralized Areas A", and the north side, which they regard as part of the Westend, one of Frankfurt's most preferred office locations (Müller International 1990: 31; Jones Lang Wootton 1990a: 12). Within this section there are three office towers more than 100 m high, an indication that this section is well established as a central location for offices.

The study on the Mainzer Landstraße by Speer suggested two additional office towers in Section 1, one on a site where the BfG-Bank owned a 20-year-old, 11-storey building, the other on the site of a multistorey car park, which was owned by the municipal Frankfurter Aufbau AG (FAAG). The Westend local plan of 1978 restricted office development and therefore a new local plan had to be set up to allow for the intended office development.

In 1986 the city planning officer presented two investors for the projects. First, the BfG-Bank intended to redevelop its site by replacing its existing office building with a 172 m tower with a gross floorspace of 85,000 m². Secondly, on the car park site the DG-Bank, in co-operation with the Dutch pension funds PGGM, planned a complex of buildings of various heights, one of 208 m, with 77,000 m² of floorspace.

This announcement was followed by some activity on the land market, because it was expected that the new local plan would give more scope for office development. Another motive for buying residential buildings around the sites designated for office towers was the Hessische Bauordnung (building code). According to the regulations in this code, a new development depends on the consent of those neighbouring house-owners whose property lies within a specified distance. This consent means that the whole property has to be purchased by the investors at exorbitant prices. In case of the DG/PGGM project the costs for buying the neighbours' consent allegedly amounted to DM80 million.

There are two further developments on a smaller scale in Section 1. A British–Dutch company invested in a building complex of 18,000 m² of office floorspace and 5,700 m² of luxury housing. The Bayerische Vereinsbank has redeveloped a site with a 10-storey building for its Frankfurt branch (see Map 20).

Section 2 has a mixed structure of prewar housing and five- or six-storey office buildings that accommodate industrial headquarters or miscellaneous medium-size service firms. In between there are small industrial units, retailers and car dealers. Some minor redevelopments or refurbishment schemes have been carried out in recent years.

Situated between two huge railway areas (the main passenger terminal to the south and the main freight terminal to the north) and a declining industrial area, Section 2 shows less preferable conditions for office development than Section 1 (see Map 19). The banking district is about 2 km away, so the location lacks the proximity advantage of Section 1. As a consequence, the image of Section 2 is not as good as that of Section 1.

On the other hand there is some potential for upgrading because of:
○ proximity to the trade fair area;
○ a fast road link to the airport, to be opened in 1992;
○ good public transport facilities;
○ schemes to redevelop parts of the industrial area as a business park, largely for offices – the so-called Galluspark;
○ development pressure in central locations which is overspilling and resulting in a centralization of adjacent areas and;
○ the policy of the city council to upgrade the area.

Nevertheless Section 2 needed an initial push to gain reputation among investors and those seeking office space. Speer suggested that two office towers should be built at the Güterplatz and the Galluswarte (Map 20). They were meant to be architectural landmarks and would attract further high-quality office development. To assist the promotion of Section 2, the planning department prepared schemes to improve the design of the street scene in the Mainzer Landstraße. Furthermore, the city council was actively trying to divert development pressure to the western part of the Mainzer Landstraße in negotiations with investors that intended to invest in other parts of Frankfurt.

In 1989 it became known that a Swedish-led multinational investor had acquired a site adjacent to the Güterplatz, the so-called Telenorma-site, at a price of DM22,000 per m^2. This was one of the highest prices ever paid for a redevelopment site in Frankfurt (Map 19). This purchase was considered to be highly speculative, as the scale of use the city council would permit was uncertain.

Because of the good prospects for upgrading the area, the real-estate fund DGI, a branch of the Deutsche Bank, acquired a block of prewar residential buildings adjacent to the Güterplatz. It is planned to redevelop the site with a mix of offices and luxury apartments. The DGI calculates that the return from this office and flat mixture will be sufficient, without exceeding the present scale of use by developing a high-rise building.

Some sites were occupied by the Deutsche Bundesbahn (the federal rail service) in the area between Mainzer Landstraße and the trade-fair area (Map 20), but the federal government instructed it to sell off those sites that are no longer needed for railway operations. The Deutsche Bundesbahn therefore sold the site of its headquarters at DM33,000 per m^2 to the Dutch pension fund PGGM. The site will be redeveloped with two office towers providing 77,000 m^2 of floorspace. The Deutsche Bundesbahn will build a new headquarters on the site of a former rolling-stock maintenance plant adjacent to the main freight terminal.

As a result of the changing attitude towards the location, which led to growing activities on the land market, average land prices per square metre, as reported by the Gutachterausschuß für Grundstückswerte rose from DM800 in 1980 and DM2,000 in 1988 to DM15,700 in 1990. The particularly sharp rise between 1988 and 1990 reflects the enhanced reputation of the of the location, which is now considered a prime area (Map 19).

Actors and their instruments In the development of the corridor along the Mainzer Landstraße, actors with a political background included the city council and the local citizens' committee AG Westend. Actors with an economic background included national and international investors; real estate service firms such as real estate consultants, architects, developers and builders; and those seeking office space.

Since the discontent over the restructuring of the residential area of Westend as an extension of Frankfurt's central business district, the public in Frankfurt have become highly sensitive to the impact of office development on residential areas. Local people fear the disadvantages of such development, such as the conversion of flats to offices, loss of affordable housing by rising rents and gentrification, traffic generation because of the concentration of jobs, and environmental impact of high-rise office buildings.

The city council had to take this into account when implementing its policies on the Mainzer Landstraße. In the case of Section 1, the city council and the investors had the same interest: to achieve a high density of offices in prestigious office towers. The council therefore used its planning instruments to support the investors by creating the necessary new planning framework. Within 15 months the statutory planning process for a new local plan to replace parts of the restrictive 1978 local plan was nearly complete, which is extremely fast. The average time to set up a local plan is normally between 36 and 48 months. Additionally, the council put the Umlandverband Frankfurt under pressure to pass the necessary amendments to the structure plan, although this was contrary to that body's planning policies.

In the case of the DG-Bank/PGGM project on the car-park site, the city council was directly involved through the municipal landowner FAAG. The FAAG was ordered to bring forward plans for redevelopment in cooperation

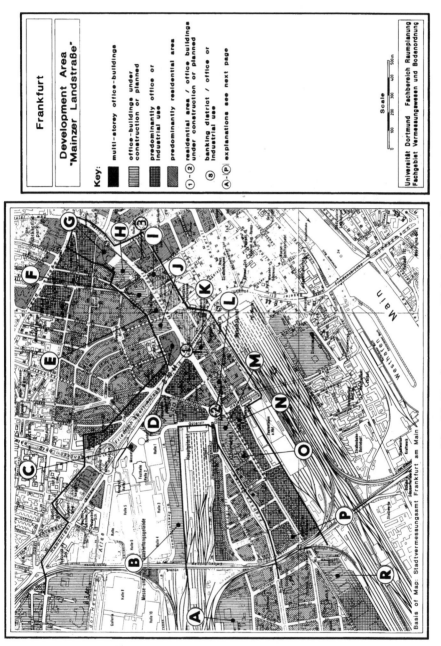

Map 20 Development area: Mainzer Landstraße.

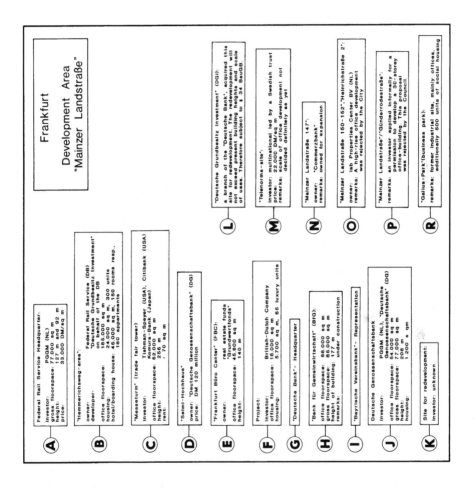

Frankfurt

Development Area "Mainzer Landstraße"

(A) **Federal Rail Service Headquarter:**
Investor: PGGM (NL)
gross floorspace: 77.000 sq m
height: 127 m and 92 m
price: 33.000 DM/sq m

(B) **"Hemmerichsweg-area":**
owner: Federal Rail Service (DB)
developer: "Deutsche Grundbesitz Investment" on behalf of the DB
office floorspace: 185.000 sq m, 300 units
housing: 24.000 sq m,
hotel/boarding house: 46.000 sq m, 150 rooms resp., 180 apartments

(C) **"Messeturm" (trade fair tower)**
Investor: Tishman-Speyer (USA), Citibank (USA), Komura-Bank (Japan)
office floorspace: 62.000 sq m
height: 256 m
rent: , 70 sq m

(D) **"Selmi-Hochhaus"**
owner: "Deutsche Genossenschaftsbank" (DG)
price: DM 120 million

(E) **"Frankfurt Büro Center" (FBC):**
owner: real estate funds "Grundwertfonds"
office floorspace: 45.600 sq m
height: 140 m

(F) **Project:**
Investor: British-Dutch Company
office floorspace: 18.000 sq m
housing: 5.700 sq m, 65 luxury units

(G) **"Deutsche Bank"- Headquarter**

(H) **"Bank für Gemeinwirtschaft" (BfG):**
office floorspace: 68.000 sq m
gross floorspace: 85.000 sq m
height of building: 177 m
remarks: under construction

(I) **"Bayrische Vereinsbank"- Representation**

(J) **"Deutsche Genossenschaftsbank**
Investor: PGGM (NL), "Deutsche Genossenschaftsbank" (DG)
office floorspace: 54.000 sq m
gross floorspace: 77.000 sq m
height: 208 m
housing: 1.200 . qm

(K) **Site for redevelopment:**
Investor: unknown

(L) **"Deutsche Grundbesitz Investment" (DGI):**
a branch of the "Deutsche Bank", acquired this site for redevelopment. The redevelopment will not exceed present building heights and scale of uses. Therefore subject to § 34 BauGB

(M) **"Telenorma-site":**
Investor: multinational led by a Swedish trust
price: 22.000 DM/sq.m
remarks: scale of office development not decided definitely as yet

(N) **"Mainzer Landstraße 147":**
owner: "Commerzbank"
remarks: owned for expansion

(O) **"Mainzer Landstraße 150-152"/"Heinrichstraße 2":**
owner: lab Properties Center BV (NL)
remarks: A high-rise office development was rejected by the City

(P) **"Mainzer Landstraße"/"Günderrodestraße":**
remarks: an investor applied informally for a permission to develop a 30-storey office-building. This proposal was rejected by the Council.

(R) **"Gallus-Park"(business park):**
remarks: former industrial site, mainly offices, additionally 500 units of social housing

with a private investor. In autumn 1987 the FAAG sold the car-park site for DM120 million to the DG-Bank/PGGM joint venture. The city council agreed to give planning permission for the office tower by July 1990, otherwise compensation of DM16 million would have had to be paid to the investors.

The formal decision on the preparation of a new local plan was passed in January 1988. In order to win the consent of residents, the city council instructed a public relations agency to design a campaign. Nevertheless, protests were made by the local residents' committee AG Westend immediately after the intentions of the city council and the investors became publicly known in 1986. Hundreds of objections to the new local plan were put forward by local people and the AG Westend succeeded in raising the issue in the campaigns for the 1989 local elections. The statutory procedure of plan preparation was not completed when, in March 1989, shortly before election day, the council issued partial planning permissions based on §33 BauGB.

The winning parties promised to follow a revised planning policy and to pay more attention to the interests of residents when considering further applications for office developments. In spite of their announcement, they failed to recall the planning permissions for the office towers because of the threat of compensation claims by the investors.

It is, however, exceptional in Frankfurt for a local plan to be prepared for office development within a built-up area. Usually the council decides planning applications affecting these areas according to §34 BauGB. This means that no public participation or consultation with other authorities, for example the Umlandverband Frankfurt, is required, since the complicated procedure of preparing a local plan is avoided. Local plans are mainly set up for the development of greenfield land and for the redevelopment of old commercial areas. Furthermore, under the conditions of §34 BauGB, decisions can be made on a day-to-day basis, which is more comfortable for office developers. In those areas along the Mainzer Landstraße that are subject to §34 BauGB, permission for office development is a matter of negotiation between the city council and investors.

The restructuring process of Section 2 is being watched with growing concern by the local residents' committee Buntes Bündnis Gallus. The committee's objectives are:

○ to achieve acceptable resettlement conditions for those who will have to leave their flats because of rising rents and to prevent harsh methods of expelling residents;
○ to prevent the gentrification of the housing stock;
○ to slow down the restructuring process to gain time for social measures and;
○ to achieve a traffic regulation scheme that keeps office-related traffic away from residential areas.

In reacting to the campaign of the committee the city planning officer promised to prepare an Erhaltungssatzung (a conservation area plan) according to §172 BauGB, but could not give much hope that rent rises as a consequence of the upgrading of the Mainzer Landstraße could be prevented. How long the city council require for the preparation of the Erhaltungssatzung is crucial if this instrument is to have any effects on the restructuring process. The Erhaltungssatzung is preferred to setting up a local plan because the statutory procedure is shorter.

Within its present planning policies the city council tries to balance out the pressure for further office development and the interests of the local people. In doing so it faces the problem that because of the highly dynamic structural changes it is unable to guide but can only react. The investors are those who set the pace of development. The planning instruments, however, are not only designed to react to, but, in the ideal case, to anticipate development. There is the problem of a time lag between the decision to set up a plan, the representation of interests in the plan and finally the implementation. Another problem is the threat of compensation claims by private investors if the city council tries to revise its planning policies totally.

There is an underlying dilemma that on the one hand the city council has to support the expansion of the service sector, especially financial services, in order to improve Frankfurt's position within the hierarchy of world cities, while at the same time it has to show some commitment to the interests of residents of areas undergoing a restructuring process. The inherent contradiction of the planning system is between an urban development largely determined by free-market forces on the basis of private landownership and the necessity to represent public interests in this process. The problem of the compensation claims of private landowners and the lack of planning-gain regulations reflects this contradiction. In the case of the Mainzer Landstraße exorbitant planning gains are evident.

The announcement of shifts in the city council's planning policies following the 1989 local elections caused some concern among the actors in the business community. There were some complaints about the unreliability of the planning framework and delays in deciding planning applications. Meanwhile this uncertainty has been replaced by confidence that present planning policies are not too restrictive. Experts agree that it is common sense that a stable planning framework should exist to enable investors to calculate the risks and profits of their developments.

Investors and those seeking office space have some instruments to prevent the city council from making its planning policies too restrictive. Besides compensation claims, the council has to be aware of the danger that capital and headquarters functions may be attracted to other European cities. Furthermore there have been incidents when the land the council required for projects was in the hands of investors of controversial office developments. Therefore the council had to enter into a bargaining process.

A range of actors are involved in the different functions in the development process. There is a trend towards a rising share of international investors. By the flow of international capital the office market in Frankfurt is directly linked to the oscillations of different national economies and the state of their property markets.

The services provided for real-estate investment and the location of firms have diversified. The real-estate consultants have extended their activities from merely providing guidance for investors in office property and those seeking office space to financing, marketing, property management, project management and research on real-estate issues. They follow an "Anglo-American" style of business, which is based on a policy of generating market interest. This business policy was introduced by real-estate consultants that operate worldwide. German real-estate agents had to adapt or lose the market for prime office locations. In addition, international investors only accept consultants of international reputation.

Some building enterprises have also enlarged their range of services. They provide comprehensive development services that include the identification of sites for office development, contact with local authorities, project design and marketing.

In prime locations, joint ventures by international investors are very important. They are able to invest large sums on the office market, but rely on the expertise of real-estate consultants or developers to place it. Furthermore, only insiders that have operated in Frankfurt for a long time are able to provide the personal contacts with the city council that facilitate negotiation of planning consents.

Thus consultants and developers are an intermediate agency between the political and the economic side of the market and facilitate the opening of the office market in Frankfurt to international investors. In the case of the Mainzer Landstraße this function is decisive for the restructuring process, especially in Section 2. Consultants have to assess the economic potential of the location and the policy framework before they are able to recommend it for investment. They therefore promote this location to get it accepted among their customers, both investors and demanders. Just as in the case of the marketing of a single project, when it is important to get a major renowned company as a first tenant, so the promotion of Section 2 of the Mainzer Landstraße will be a success when the first investor develops a prestigious project. The more investors and customers of repute that can be interested in this section, the more will follow.

The activities of renowned architects are also important. With their concepts and visionary drafts for office development they bring a location or site to people's attention. Architecture is very important to create an image that is attractive to those seeking offices. In case of the Telenorma site, the investors tried to overcome the reluctance of the city council to permit a high-rise building by instructing H. Jahn, architect of the Messeturm (trade-fair tower), to design a 200 m double tower. This scheme was presented to the city council and the public as an example of how the appearance of the area could be improved.

Those seeking office space are the unknown quantity in project calculations. They normally prefer established locations, because their company image is often linked to their addresses. The question for investors and their consultants is whether a new location will attract sufficient demand to pay off the investment. In Section 2, land prices have risen faster than rents have followed because of the usual five-year rent contracts. This is because the subject of a land purchase is not the existing use of a site and the return from an existing building but the return on its future use. Therefore office development has to be calculated on the basis of the expectation of future rent rises. The market process will show whether these rents can be realized.

The city council's task is to set priorities between the different interests of investors and residents. At the moment it looks as if the council, whatever its attitude towards the social consequences of office development, has come too late with the implementation of a planning framework that is balanced between development pressure and social and environmental issues. But there are signs that at least some real-estate consultants see it as a responsibility of investors to support the city council in its introduction of a balanced development policy. They are aware of the danger of one-sided policy in favour of office development that neglects social and environmental dimensions.

Poseidon Haus – an office-building development

This section describes the development of office space in the Poseidon Haus mainly from the investor's and developer's point of view. It is based on information provided by the company that manages the project.

The Poseidon Haus is situated on a $6,800 \text{ m}^2$ site opposite the main entrance to the trade-fair area (Map 20). Before the development the site was owned by the city of Frankfurt and used as a car park. In 1981, the city approached a project management company (PMC) with the suggestion to develop the site with offices. The management company was able to present one of its clients as investor for the project. Three months after the first contact between the city and PMC the investor, the Dutch pension fund PGGM, purchased the site for DM35 million, about DM5,150 per m^2. This price was about 30% more than average price for prime locations as reported by the land valuation committee (Map 19). Although this might indicate that the city got a preferential price, rises in land values in subsequent years made the investment profitable for the PGGM. At that time the office market was recovering from the depression and oversupply crisis during the mid-to-late 1970s.

The planning background was quite unusual for an inner-city location in Frankfurt. There was a local plan, dating from 1969, that allocated the site for commercial use. The plan provided for more than 60% of the site to be built on at a floorspace ratio of 4.5. Thus the investor could claim a right to build according to the local plan.

A planning team was set up chaired by PMC, which included an architect, several building engineers with special responsibilities, a landscape designer and several other subcontractors. Drafts for the building were discussed and amended in weekly meetings. In autumn 1981 a planning application was made. The first partial planning permission was issued in spring 1983. Works on the building started almost immediately (Fig. 11.3). The final draft showed a building complex with three parts with 17, 7 and 5 floors. The net floorspace was 24,800 m² with 20,100 m² used as offices.

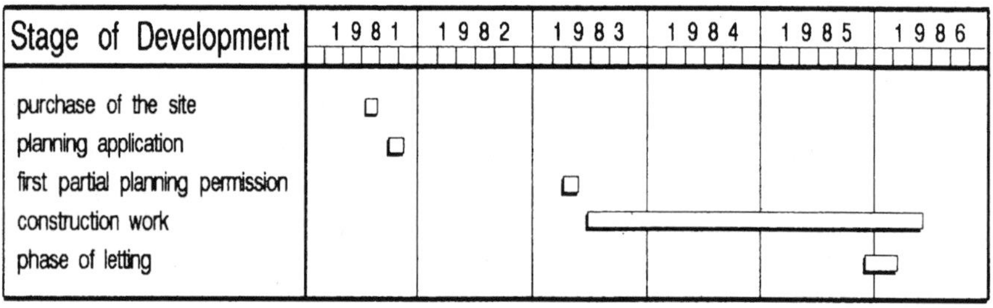

Stage of Development	1981	1982	1983	1984	1985	1986
purchase of the site	□					
planning application	□					
first partial planning permission			□			
construction work						
phase of letting					□	

Figure 11.3 Timetable: office project Poseidon Haus.

While the council was considering the application, PMC tried to negotiate a higher scale of use, exceeding the local plan allocations. According to the company this was not for financial reasons but for architectural ones. In turn the investor was prepared to increase the price for the site. The council, however, turned down the higher scale of building. It is our interpretation that the council was not willing to risk public discontent by going against its local plan. While waiting for the decision PMC visited the building control department weekly. Such regular contact with the public authority is essential for developers in order to shorten the authority's decision-making process.

With regard to the investor, who was seeking medium- and long-term investment opportunities, the project was built for the rented office market. The building was designed to be attractive to most types of users. Thus the internal structure of the building was kept as flexible as possible, allowing both for small rooms and large office units. Another reason for having a flexible structure was to minimize effort when the building or parts of it have to be adapted to changing requirements of users. Furthermore, the reselling of the building is easier if it has a flexible interior structure.

Building costs were calculated at DM62.5 million. The real costs amounted to DM61 million, and the price of the land was DM36.5 million. The total costs of the project, including interim financing costs, were DM125 million. The project was self-financed by PGGM. A special company was set up for the project that obtained its capital stock from PGGM, to which it had to pay interest.

Marketing of the project began when the shell of the building was completed. This involved not only potential users but also the general public. An important factor in the marketing concept was the name of the building. The name Poseidon Haus was launched as a trademark in newspapers and magazines. Through a public relations campaign based on the name and the exterior appearance of the building the attention of potential users was drawn to the project.

In 1986, about six months before the completion of the building the first $4,000 \, m^2$ were let. Three months before completion, the remaining floorspace was completely let to Dresdner Bank AG.

According to PMC, the completion of the project was never in doubt. The greatest difficulty was that building work was faster than the planning consent procedure of the authority. This project, PMC points out, cannot be seen as typical for office development in its experience. It was less difficult to achieve, particularly because no consent had to be obtained from neighbouring house-owners, because of the old local plan. In general PMC denies that projects can be compared with each other at all, because each has its special planning situation and individual architectural environments.

11.3 Köln

Introduction

As research projects have shown, larger German cities are showing a trend towards migration back into the urban area (Dangschat & Blasius 1990). This process of reurbanization is carried out by distinct categories of household and households at specific income levels. One-person households and new types of housing communities, often with above-average incomes, have rediscovered the inner city as a new and convenient residential location. Yet not all inner-urban areas are affected by this trend. Demand is particularly strong for high-quality housing neighbourhoods built around the turn of the century.

On the German housing market the process of gentrification involves modernization of houses in traditionally working-class areas of a city, conversion of rented flats into apartments, and a change in the social structure of the particular neighbourhood (Kreibich 1990).

Linked to this is an increase in the prices of land, rents and property. A different spectrum of actors in the property market emerges and different attempts to control the market are made by the national, regional and local governments.

The following case study describes the requirements, practice and effects of the ongoing trend of *Umwandlung* (conversion). The term conversion covers all flats that were originally built as rented units and were later transformed into apartments for sale (Hüttenrauch and Schaar 1990).

The process of conversion of flats is currently under much discussion in Germany. The continuing loss of low-rent housing for low-income households and the tendency for social segregation associated with conversion to apartments, is subject to many complaints but an end to the process is currently not in sight.

The study area in which the apartment market and the process of conversion was monitored is the *Unterbezirk* Neustadt-Süd, a statistical sub-district of the Innenstadt (inner city) (see Map 21). By statistical analysis the relationship between the number of converted flats, their time of conversion and their date of sale as apartments may be worked out. The behaviour of different actors involved in this process is also examined.

The geographical, social and economic structure Köln is located in the southwest of Nordrhein–Westfalen and is its largest city. In 1989 it had a total population of 983,454 inhabitants in 472,904 households and covered an area of 40,512 ha (see Map 15 in the Düsseldorf case study, Ch. 11.1).

After a decrease in population in the early 1980s, which reached an all-time low in 1985 with 965,274 inhabitants, there has been a constant increase since 1986. The largest foreign ethnic group is the Turks (65,847 citizens), followed by the Italians (19,902 citizens).

Köln is at the core of an agglomeration that encompasses the sovereign cities of Leverkusen (157,358 inhabitants in 1988), Bonn (282,190 inhabitants in 1988) and Aachen (233,255 inhabitants in 1988) as well as the surrounding counties (Erftkreis, Rhein–Sieg Kreis, Rheinisch–Bergischer Kreis) with an overall population of about 2,295,512 in 1988. Thus nearly a quarter of the region's population live in Köln.

Within a radius of 80 km there are 28 *Großstädte* (cities with a population of more than 100,000 inhabitants) including the Ruhrgebiet industrial agglomeration and the state capital of Düsseldorf (see Ch. 11.1), with which Köln has historically been in a state of competition or rivalry. The city of Köln belongs to the European Region district (EC) No. D52 and is called Region Köln.

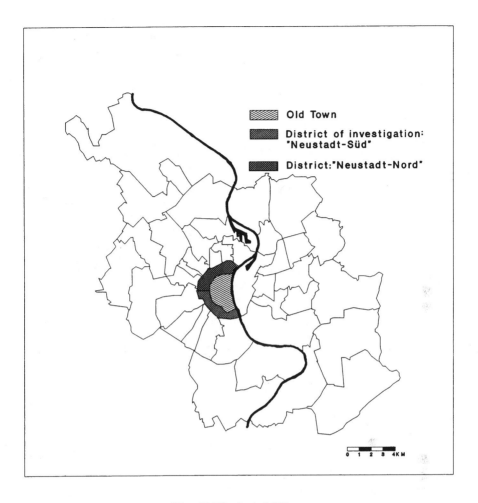

Legend:
- ▨ Old Town
- ■ District of investigation: "Neustadt-Süd"
- ▦ District: "Neustadt-Nord"

Scale: 0 1 2 3 4KM

Map 21 Districts of Köln.

Today Köln has a broadly based economy, giving the city the strength to overcome national or global weaknesses in certain industrial sectors. Industrial sectors with national importance in Köln include car manufacturing (Ford employs 28% of the industrial labour force), mechanical engineering (18.7%), chemicals (Bayer, BASF, Shell etc, 13.3%), the electro-technical industry (12.7%), coal-mining (Rheinbraun) and other industries (27.3%). Largely as a result of technological development, about one-third of the jobs in the industrial sectors have disappeared during the past 17 years, while during the same period the number of service-sector jobs has nearly doubled, reaching 116,602 at the end of 1987.

Although many experts do not consider Köln to be a booming industrial centre, some real-estate agents feel that it has enormous potential, especially with the imminent establishment of the European Single Market (Aengevelt 1990c).

Public policies for future development The Flächennutzungsplan identifies 310 ha of land for housing for the whole area of the city that will be available in about 10 years.

In 1978 a comprehensive development plan, subsequently amended in 1989, was passed by the city council. This defines the general development goals for the inner city and the surrounding Neustadt area. In particular, it identified policies relating to road planning and improvement of living conditions in the housing areas. One main aim was the stabilization of inner-city districts for residential use.

Economically, the municipality is trying to regenerate the inner-city business districts. The prime example is an attempt to increase its image as a long-standing media-centre of national importance. The Media-Park, a 20 ha prestigious commercial development with approximately 160,000 m² of office, studio and administration space at the northwestern edge of the Neustadt, is currently under construction. It completely changes the scenery in this part of the city, since the area was previously a run-down rail freight station.

The housing market in Köln and the Innenstadt district

The rented housing market In general, the housing market is determined by the rented market. The 1987 census counted about 370,000 (82.6%) out of all 450,000 flats as rented. With a margin of error of 3–5% because of people living in sub-tenancy of some kind, approximately 78% of the 481,392 households in 1991 were living as tenants. In 1991 in the Innenstadt, 69,248 (93.8%) of flats were rented, out of a total of 73,824 flats with 137,409 residents.

Free-market housing rents have exploded during the past four years because of the current shortage (see Fig. 11.4), while subsidized social rents still remain at a level of DM6.70–7.00 per m²on average.

In 1987 116,386 units (31.2%) of the 373,019 rented flats were publicly subsidized through the 1. Förderungsweg of the social housing programme according to the II. Wohnungsbaugesetz. In 1989 the head of the Köln Wohnungsamt (board of housing) stated that an estimated 104,000 flats remained in public ownership in the city. The city could name the tenant in only 40% of the cases because of the given contracts. In 1985 there had been 6,281 city arrangements for flats, compared with 22,155 people demanding them. In 1989 there had been only 2,985 arrangements, compared to 24,200 demanded.

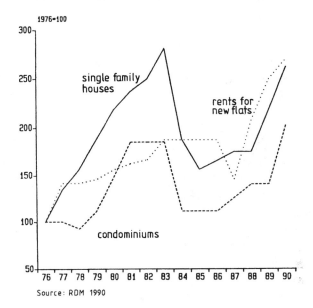

Figure 11.4 Development of rents compared to prices for single family houses and condominiums from 1976 to 1990.

The discrepancy between demand and supply in the social-housing sector is expected to increase dramatically within the next 10 years. Until 2000, more than half of the remaining social housing stock will cease to be in public ownership. This amounts to an annual decrease of 3,000 units.

Although the social rented housing market is important, especially for low-income groups, 69% city-wide and as many as 85% of all rented flats in Neustadt-Süd are privately rented and are not in public ownership. The only reason for intervening on this market can be in cases of rent usury.

According to §2 Miethöhegesetz (the federal rent limitation statute) the landlord can raise the rent by 30% over three years until it has reached the local *Vergleichsmiete* level (average local comparable rent). This Vergleichsmiete is rapidly rising with the effect that the possibility of capital gain for landlords increases substantially and investing in housing is expected to be profitable in the mid to long term.

The apartment market In Köln there are about 36,000 apartments, compared with a total of approximately 450,000 flats (i.e. only about 8%). Of these, about 11,700 are owner-occupied. In the Innenstadt district there are about 7,500 apartments within the total of 74,000 flats (i.e. 10%). Of these, 1,900 apartments are owner-occupied (Stadt Köln 1989).

239

Average purchase prices for apartments are currently about DM3,800 per m² for new buildings. In the case-study area of Neustadt-Süd, prices range from DM4,500 to DM5,000 per m². Between DM3,500 and DM4,000 has to be paid in Köln itself for a second-hand apartment, while the prices drop to DM2,000–2,500 for a rented flat. Penthouses in prime locations reach prices of up to DM10,000 per m².

The city Wohnungsamt (board of housing) identifies the following as actors in the apartment market:

○ private persons and tenants;
○ non-profit-orientated housing companies;
○ profit-orientated housing companies such as subsidiaries of insurance companies, banks and industrial groups;
○ other entrepreneurs such as real-estate agents, developers and professional converters;
○ other owners such as the state, city, Church and charity organizations.

There are various reasons for the change in the ownership structure of flat complexes. As mentioned above, a very high percentage of flats are still rented. Many of them are owned by profit and non-profit housing companies. Since the possibility of capital gain on the housing market is expected to increase, smaller companies (often professional converters) and private persons are currently entering the scene.

As studies have shown, many professional actors and investors changed to the office-space and business-park sector after the market crash in 1983–6. However, some of them have shifted back into the housing market, although very gradually. Furthermore there is a clear trend towards owner-occupation of apartments. Professionals in the Köln housing market have noticed a tendency among clients to purchase an apartment rather than spend more of their total income on rent.

Public policies on the apartment market

In terms of public policy, the apartment market is no longer an important area of responsibility for the municipality. There are no clear policy statements about the conversion of rented flats into apartments except where the conversion constitutes a "luxury modernization". Since the authority is committed to the creation of equal living conditions in its area of responsibility, it continues to attempt to prevent spatial and social disparities.

A declared goal of the city administration is to maintain the inner city as an attractive housing area in order to avoid problems of slum development and dereliction. In the 1989 *Entwicklungskonzept Innenstadt* (the development concept of the inner city) of 1989 the city noted that housing in the core area was endangered by pressure on the predominant housing land-use from other functions, partial neglect of the housing stock, decrease in the availability of low-rent housing units because of luxury modernization, continuing loss of neighbourhood green and open space, increased impact of traffic through lack of parking space and noise pollution, and loss of neighbourhood identity.

The planning department designated Rahmenplanungsgebiete (areas with non-binding development plans), Stadterneuerungsgebiete (urban renewal areas) and Modernisierungsschwerpunkte (centres of improvement) as guidelines for public planning and private investments. Furthermore the City Council has declared Erhaltungsgebiete (zones with statutes of preservation, according to §172 ff BauGB) and Sanierungsgebiete (formal urban renewal zones, according to §136 ff BauGB) with legally binding status.

The local administration distinguishes two general types of urban renewal areas, dealt with by different strategic planning instruments (Schulz 1990: 185):

(a) Problematic areas with pressure for transformation and high investment potential and,
(b) Problematic areas in disadvantageous locations with less pressure for transformation.

In the case of type (a) areas, the city tries to reduce the development expectations and slow the transformation process mainly by declaring formal Sanierungsgebiete (urban renewal zones) according to §136 BauGB. This has the consequence that all sales, private construction measures and divisions of property have to be approved by the local authority. The loss of low-rent housing through luxury modernization can best be prevented in formal urban renewal zones (Schulz 1990: 185). A formal renewal zone, for example, has hampered attempts at transformation by private investors in the Stollwerk-viertel. The Erhaltungssatzung (a statute of preservation) is the important instrument that is often used to control development.

In the case of type (b) areas, the city intervenes basically in order to generate market powers. Such measures include Wohnumfeldverbesserung (residential area improvement) by means of greening streets, traffic-calming and better public transport links to that particular part of the city. Also the instruments mentioned in type (a) areas (Sanierungsgebiet, Erhaltungsgebiet) are used for public interventions. A special focus has to be on the formal urban renewal zone, because this instrument enables the local authority as well as the property-owners to take advantage of state and federal funding.

Umwandlung (conversion) in the Neustadt-Süd

This section evaluates the importance, as well as the current and future share, of the process of conversion of formerly rented flats into apartments on the housing market for the Neustadt-Süd. It is based on the thesis that, by merging data on the "certificates of completion" and on actual sales in the district, one should be able to detect connections and trends. Furthermore, the locations of cases have been identified, and the relationship between their concentration and public planning measures (for example designated urban renewal zones) was identified. A flat that was formerly rented is defined as having been converted once it has been sold as an apartment.

Description of the subdistrict of Neustadt-Süd The subdistrict of Neustadt-Süd is part of the so called Köln Neustadt, a city expansion planned by the architect H. J. Stübben at the end of the 19th century. The district is subdivided into a north and a south part (see Map 21).

The Neustadt-Süd is bounded in the north by the Lütticher Straße, in the northeast to the east by the Innerer Wallring (Mauritiuswall, Pantaleonswall, Kartäuserwall, Severinswall), in the east by the river Rhein and from the south to the west by the railway line (Map 22). It encompasses 260 ha, which is 17% of the area of the Innenstadt. In 1991, 39,436 inhabitants lived in about 23,000 flats, of which 89% were rented. There are about 2,500 flats (11%) that have been converted to apartments, and approximately 500 are owner-occupied (Landesamt für Datenverarbeitung und Statistik 1987). Flats have an average size of 61 m². With 151.7 persons per hectare the Neustadt-Süd has the highest density of all housing subdistricts in Köln.

In the Neustadt-Süd, 46% of the houses were built between 1949 and 1986. The case-study area has the largest number of residential buildings built before 1918 in the city (30% of all residential buildings in Neustadt-Süd). In the whole Innenstadt district, only 8.5% are of this age. Such buildings are usually five or six storeys high. The local authority has estimated a potential to develop 4,000 apartments by the use of vacant housing land and currently vacant floor space.

Local policies for the Neustadt-Süd The above-mentioned Entwicklungskonzept Innenstadt from 1989 defines 26 development goals with housing as the predominant land-use in the Neustadt-Süd. Along the streets as well as along the main traffic arteries entering the inner city, ground-floor street frontage retail and service enterprises are allowed. Along the Hohenstaufenring and the Pfälzer Straße on both sides and south of the Sachsenring, land is allocated for office development. According to the *Entwicklungskonzept Hauptverkehrsnetz* (the development concept for traffic), public transport on rail and road links the Neustadt with the core of the city and the outskirts. Three major transfer points are to be installed between bus, underground and tram along the border between Neustadt and the city centre. It is hoped that the problem of rush-hour traffic will be minimized by increasing the attractiveness of public transport to commuters.

In the case-study area, as shown on Map 22, the Wohnbereiche (housing areas) No. 6 Zülpicher Straße/Rathenauviertel (14,700 inhabitants), No. 10 Südliche Neustadt (17,100 inhabitants) and the southern part of the Wohnbereich No. 9 Pantaleons/Pfälzer Viertel (4,500 inhabitants) are areas where urban improvement programmes including the greening of facades, new street furniture, tree-planting, street-lighting and traffic-reduction measures are planned.

The predominant part of Wohnbereich No. 6 is defined as Wohnbereich mit besonderem Wohnbestandsschutz (an area with a particular protection with regard to the residential use) and as a Rahmenplanungsgebiet (development plan area). Wohnbereich No. 10 comprises parts of a formal urban renewal zone, and is covered almost in total by a development plan area, including three major areas reserved for housing.

The process of conversion The process of conversion begins when the owner of a building containing rented flats intends to establish separate flat ownerships. The legal process of such conversions consists of two steps. First, the owner needs an Abgeschlossenheitsbescheinigung (certificate of completion – CoC) from the Bauaufsichtsbehörde (local building authority). The CoC defines each flat as one single, separate unit. Therefore the technical standards of noise and fire protection must be met for each flat. The CoC is always necessary for conversion. Secondly, the owner has to give a Teilungserklärung according to §8 Wohnungseigentumsgesetz (declaration of division) for entry into the Grundbuch. A property of, for example, 10 rented flats is recorded from this point on in the land register as 10 separate properties. Afterwards these separate flats can be sold as apartments. As a result the purchaser can cancel the contract of tenancy with the original tenant, but has to take into account the period of notice fixed by rent law.

The number of CoCs (step 1) does not necessarily monitor the real number of sales (step 2) and thus of finalized conversions. There is no legal rule connecting the two steps and in practice the actors often make use of this. This is because property-owners often consider the CoC to be an option that increases the sales potential of their property, and therefore its value. They may then be well placed to wait for more favourable market conditions such as an increase in demand or prices.

Meanwhile, conversion has been hampered by recent court decisions that have ruled that technical standards for conversions should be based on standards applicable to new dwellings and no longer on standards applicable when the property was built. This has particularly hit the formerly attractive conversion market, especially for premises constructed before World War I, and the number of applications for CoCs has recently declined.

Figure 11.5 demonstrates typical chains of participants in the process of conversion. A few building-owners are replaced by many single flat-owners.

This can, for example, be caused by the attempt of the old owner to realize capital, by the death of the former long-term owner, or by an heir not being interested in keeping the property. Occupants can be faced with the following situation. They can keep their contracts of tenancy, which are secured by tenants' protection law, but with the risk of rising rents. Alternatively they can purchase their flats for owner-occupation. In the case of a radical professional conversion, however, they can be pushed out of the flats they occupy.

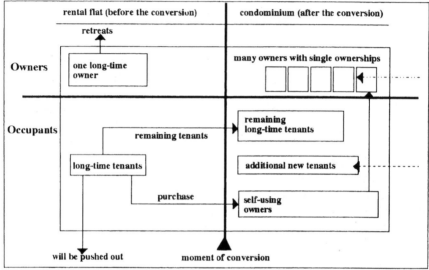

Source : Kreibich,B. et. al.: Gutachten zum Wohnungsgesamtplan Köln, Grundszenario zur Wohnungsmarktentwicklung, Dortmund 1989, p. 103.

Figure 11.5 Actors in the process of conversion.

Conversions are often carried out by professional companies and private individuals who systematically buy old houses or even complete blocks for sale after modernization as apartments, often making a large profit.

In general, a change in the social structure of a converted housing complex can be noticed. This is evident, for example, from the continuing retreat of foreign citizens, one example of a marginal social group, from Köln-Neustadt.

Research and results

The correlation between the number of CoCs and the actual sales as apartments should be examined, both numerically and over time. Their locational concentration and their dependence on public policy has to be taken into account. Three stages of investigation were carried out:

1. Testing how typical the district Neustadt is, to evaluate whether the findings could be applied to other Köln inner-city housing districts by:
   ○ evaluating the CoCs in different subdistricts of the Innenstadt (inner city) according to different kinds of actors. The statistics of the Wohnungsamt (board of housing) could be used. These record the applied and declared CoCs since 1983 (see Fig. 11.6). In the subdistrict of Neustadt-Süd these figures have been collected since 1985.
   ○ evaluating the sales of apartments according to the Kaufpreisstatistik (statistics on sales prices) of the valuation committee from 1981–9 in Neustadt-Süd (see Table 11.3). Thus data common to both of these categories exists for the five-year period from 1985 to 1989.
2. Location of individual subjects of applications for CoCs within the area of the Neustadt-Süd in relation to the areas of public policy designations (see Map 22).
3. Merging the data on the CoCs with actual sales from the valuation committee data to evaluate the temporal and numerical interrelationships (see Tables 11.4 and 11.5).

Figure 11.6 illustrates the number of CoCs in the Innenstadt (CoC 1) as well as in the subdistricts of Neustadt-Süd (CoC 102) and Neustadt-Nord (CoC 104) (see Map 21). Additionally, the number of sales of apartments, both converted and unconverted, within the city subdistrict of Neustadt-Süd is shown. The graphs result in similar developments, thus the study area Neustadt-Süd can be seen as typical for the inner-city property market in Köln.

Comparing Figure 11.6 with Figure 11.4, it can be seen that until 1989 the curves are very similar. Although not directly connected, the level of prices, of rents and of the number of CoCs is similar, because both react to general market conditions. If the prices are going down, it seems that interest in conversion of flats decreases. The slope of CoCs since 1989 can be interpreted as a result of the court decision referred to above on the technical standard of apartments.

The number of conversions can be influenced by public intervention as well as by general market conditions. This can be shown by comparing the locations of buildings for which successful applications for CoCs have been made between 1985 and 1990 (see Map 22). Neustadt-Süd is an area of intensive municipal planning measures (formal renewal areas, areas with development plans that are not binding). The low numbers of conversion projects is obvious, especially in the areas of formal urban renewal zones (Sanierungsgebiete). These areas have been protected from professional conversions. However, Map 22 shows that there is a tendency toward conversions in the housing areas that are located immediately next to them.

Map 22 shows that non-binding development plans are not strong enough

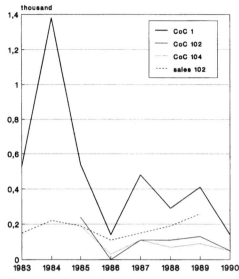

Figure 11.6 "Certificates of Completion" and sales of flats in selected districts from 1983–90.

to prevent the conversion of flats. By comparison, these areas are usually more attractive for investors because of the public measures.

Table 11.4 is a working table showing merged CoC and sales data. The two statistical sources are differently structured, and the period of the data (1985–9) is quite short in view of the time lag between a CoC and the sale of an apartment, so the results have to be handled with caution.

The results of merging the data are summarized in Table 11.5, differentiated according to the age of the buildings. This confirms that buildings built before 1924 attract the most interest for conversion. Of 584 CoCs granted, only 159 (27%) could be identified in the file of apartments sold. Nevertheless out of this small number, about 90% were sold within two years.

The results have to be related to the general market conditions in the examination period between 1985 and 1989. The decline of market prices in the mid-1980s is the main reason why only a small proportion of CoCs resulted in converted flats. It can be supposed that on the one hand most of the non-profit-orientated actors sold their flats within the short period of two years and on the other hand the professionals operated within the market cycles. They do not sell the flats because they expect prices to increase again. The professionals also usually seek to avoid speculation tax by keeping ownership for more than two years.

It is to be expected that in the current period many of the converted apartments will come on to the market in Köln because general market conditions have considerably improved and because of the accumulated stock of flats on which CoC applications have been granted.

246

Table 11.3 Results of the 'Kaufpreisstatistik' (statistics on sales prices) for 1981–1989 for the subdistrict 102 Neustadt-Süd.

Year	Total sales	Average price DM/sq m	Sales by the age of building and the average price/sq m												
			A		B		C		D		E		F	
			n	p	n	p	n	p	n	p	n	p	n	p
1981	173	2,355.3	53	1,703	7	1,348	63	1,905	13	2,379	12	2,810	25	3,986
1982	51	1,901.5	27	1,692	5	1,573	13	1,997	6	2,345	–	–	–	–
1983	153	2,155.5	54	1,635	3	1,939	67	2,052	15	2,278	5	2,691	9	2,338
1984	220	2,129.1	65	1,723	3	1,800	66	1,886	54	2,639	32	2,598	–	–
1985	194	2,169.3	83	1,796	26	1,398	34	2,089	45	2,396	1	1,902	5	3,433
1986	106	1,980	54	1,693	2	1,646	39	1,589	9	2,184	1	2,142	1	2,626
1987	147	1,735.1	68	1,730	12	1,206	43	1,436	15	1,871	4	1,480	5	2,688
1988	191	2,017.5	99	1,814	8	1,530	37	1,678	15	1,938	4	2,097	28	3,049
1989	267	2,119.4	134	1,811	12	2,289	60	2,028	29	1,909	9	2,048	23	2,631
total	1,502	2,062.5	637	1,733	78	1,637	422	1,851	201	2,215	68	2,221	96	2,964

Source: Municipality records

Notes:

Age of building

A = pre-1924　　　　　B = 1925–1948

C = 1949–1960　　　　D = 1961–1971　　　　n = number of sales

E = 1972–1980　　　　F = 1981–　　　　　　p = price per sq m

Table 11.4 Outline of the working table for merging data, property built before 1924.

Location	type of building	completed actor	completed year	completed flat units	sold year	sold flat units	CoC not sold	sold-no data on CoC	time frame	statement
Brüsseler Str 12	C	1	11/85	5	86	4	–	–	1	
					89	1	–	–	4	
Darmstadter Str 4	D	4	9/89	11	89	11	–	–	0	
Darmstädter 19	D	1	7/85	9	88	3	–	–	3	
					89	3	3	–	4	
Engelbert Str 1	D	1	4/86	16	88	3	13	–	2	
Engelbert Str 17	C	1	2/85	5	85	3	2	–	0	
Gabelsberger 11	D	1	6/90	2	81	10		10		not
					82	4		4		registered
					83	1		1		
					84	5		5		
					85	3		3	>1	
					86	2		2	>2	
					87	3		3	>3	
					88	3		3	>4	
					89	4		4	>5	
Händel Str 8	D	1	12/87	10	88	7			1	
					89	1	2		2	
Kyffhäuser Str 8	C	1	1/88	5	89	2	3	–	1	
Mainzer Str 76	D	1	12/87	9	87	1	–	–	0	insecure
					88	3	–	–	1	perhaps comp-
					89	3	2		2	pleted earlier
Mainser Str 78	D	4	5/85	1	85	19		18	>1	not
					86	2		2	>2	registered
					89	1		1	>4	
Metzer Str 35	C	1	11/85	5	85	1		1		
		1	10/88	2	86	1		–	>1	
					89	1	>4	–	>1	
Roland Str 12	F	4	10/89	21	89	6	15	–	0	
Roland Str 65	D	4	2/87	10	86	1		1	>2	not
		1	8/88	2						registered
		1	11/88	10				22		

Type of Building	Actors	Table of results		
			time-frame cases	units
A = single family	1 = private person		0 12	44
B = two families	2 = non-profit housing		1 9	24
C = 3–6 flat units	companies		2 5	9
D = 7–12 flat units	3 = housing companies		3 1	3
E = 13–20 flat units	4 = other entrepreneurs		4 3	5
F = 21 +	5 = other owners		5 0	0
			26	85

Table 11.5 Resulting table of the merging of data of the file of certificates of completion and the statistics of sales.

Type of building	registered, con- verted, sold	converted within two years %	potential: CoC – not yet sold
single family	85	90.6	79
two families	3	100	10
3–6 flat units	34	91.18	24
7–12 flat units	13	100	14
13–20 flat units	–	–	–
21+ flat units	24	58.33	65
total	159	88.15	192

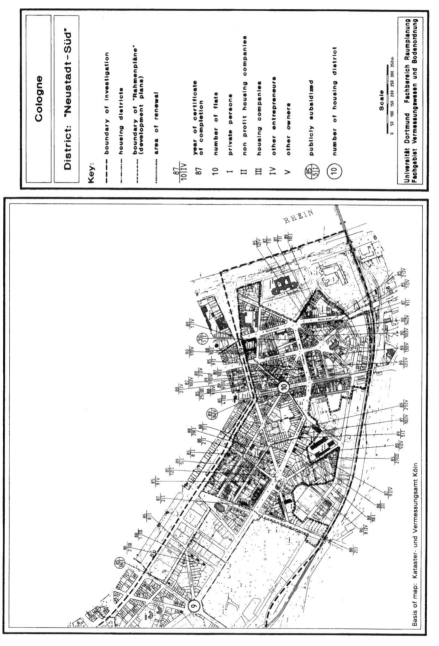

Cologne

District: "Neustadt-Süd"

Key:

—·—·— boundary of investigation

------- housing districts

·········· boundary of "Rahmenpläne" (development plans)

··········· area of renewal

$\frac{87}{10|IV}$

87   year of certificate of completion

10   number of flats

I   private persons

II   non profit housing companies

III   housing companies

IV   other entrepreneurs

V   other owners

$\frac{85}{5|I}$ publicly subsidized

⑩   number of housing district

Scale

0   50   100   150   200   250   300   350m

Universität Dortmund   Fachbereich Raumplanung
Fachgebiet Vermessungswesen und Bodenordnung

Basis of map: Kataster- und Vermessungsamt Köln

Map 22   District Neustadt Süd.

PART IV
Evaluation

CHAPTER 12

Evaluation of the functioning of the market for urban land and property

12.1 Residential use

In general, the land and property markets in Germany have functioned satisfactorily over the past few decades. However, there have been, several periods of market failure, especially during times of shortages of building land or flats in the residential sector. It is not usually the lack or inadequacy of public regulations that causes these failures. Many difficulties are attributable to ignorance about available legislation or people's inability to apply it advantageously. In addition many public regulations are sometimes counterproductive and inflexible, and hinder positive reactions to changing market situations.

It may be concluded, first, that the real problems in the market are not generally nationwide but are at the regional or local scale. Secondly, the problems mostly affect those on lower incomes. Moreover the market shows many fluctuations, with periods of housing shortage, or contrasting periods of quiet market conditions. Ownership distribution, adequate supply and access to the market and appropriate prices are not always fulfilled for all social groups in all regions. The relatively few owner-occupiers is a sign that even people with middle incomes have difficulties in fulfilling their wishes in respect of the land and property market.

The present distribution of ownership is not satisfactory, because only about 40% of all households are owner-occupiers, and the average age at which properties are bought is 40 years. In particular, young families are not able to buy their own houses, although the majority would like to. By contrast, an increasing number of single-person households in inner-city locations do not wish to become property-owners. In inner-city areas the rate of ownership continues to decline and is currently about 20%. This means that there is a concentration of ownership among relatively few people or companies in the economically important regions. Historically, German inner-city areas have always been characterized by a large stock of rented housing.

Levels of ownership are too low and ownership is too expensive for many people. Most families can afford to be property-owners only in the countryside surrounding the cities or in the more rural regions. This results in long commuting distances and continually dispersing settlement structures.

The present supply of building land is considered sufficient on a nationwide scale. There is, however, a clear shortage in the fast-growing cities, particularly in the inner areas and even in the highly populated surrounding countryside of the metropolitan areas. In many cases there is enough Bauerwartungsland zoned Flächennutzungsplänen or even Rohbauland zoned by Bebauungsplänen but still unserviced, although Baureifes Land (completely serviced land) is rarely available. There are several reasons for this. These include the powerful environmental protection policies of the regional planning authorities and city councils, increasing development costs and often the lengthy planning and implementation process. The strong position of landowners is another hindrance. Sometimes methodological problems of estimating real demand for building land over long time periods must also be taken into account. Therefore, in many cases the land market cannot react to unexpected demand as soon as is necessary.

Many problems in the residential land market are caused by the municipalities' lack of long-term land-assembly and land-banking policy, often because of inaccurate predictions of future demand, as well as limited public finances.

Hoarding of building land is another unsolved problem in the land market. The problem of supplying enough built-up land will continue, because there is no legal obligation on landowners to bring their serviced plots into use according to the land-use plans. The behaviour of landowners is supported by the taxation of land, which is not based on real market values. A revaluation according to current values and an increasing acquisition of land by the municipality as the intermediate owner would help to solve this problem. Furthermore the new Wohnungsbauerleichterungsgesetz (legislation facilitating housing) should be applied more often. The Entwicklungsmaßnahme (municipal development measure) could help to supply enough building land at acceptable prices.

Prices for building land and built properties have increased in recent years to an extraordinarily high level. The ratio between the average income level, the interest rate and the level of property prices has continually diverged, so that the opportunity for low- and middle-income groups to become property owners is reduced or even eliminated.

Access to ownership is especially difficult in growing regions, in the inner-city areas and in the highly populated surrounding countryside of the metropolitan areas. Average land prices of up to DM1,500 per m^2 (see the Stuttgart case study, Ch. 7.2) and continually rising rents may be regarded as a reason for market failure. Clearly it is not satisfactory that only a quarter of new houses are built by households with an average net monthly income below DM3,000, while the average net monthly income of all German households comes to DM2,300. A savings period of approximately 100 monthly incomes or 8–10 working years is not conducive to widespread home-ownership. On the other hand, in several growing regions an increasing number of households has been forced into ownership (mainly of apartments) because of rapidly rising rents.

The difference in prices between urban areas, the surrounding countryside and the peripheral regions results in an unfavourable allocation of land. Different land-uses, such as for residential, industrial production and office purposes, have been separated from each other, and many urban regions now have districts with an unbalanced mixture of land-uses. For example, nowadays it is more or less impossible to develop social housing in centralized areas, because land prices are too high and the limit for social rents can therefore only be maintained with unjustified massive subsidies.

12.2 Commercial use

In contrast to the housing sector, the commercial property market is in general no hindrance to economic development. For example, the land market is not responsible for unemployment. In only a very few cases has company formation or relocation failed due to the lack of suitable plots and high property values. In the five new Länder in eastern Germany, however, there is a different situation, because the distribution of ownership is still unclear, hindering economic development.

There are three main reasons for this situation. First, the commercial uses (especially in the service sector) are more profitable than others, so that they are usually the winners in the competition between different land-uses. Secondly, land and property prices are only one criterion among many for company location. Thirdly, municipalities have generally attached more importance to a long-term land-banking policy for commercial and industrial land, so they are able to offer sufficient land. The dependency on the revenues of the Gewerbesteuer force the municipalities to adopt this consequent supply policy.

In addition to this general evaluation of the market conditions there are also regional differences. Particularly in the inner areas of growing cities such as Berlin, Hamburg, München, Stuttgart, Frankfurt and Düsseldorf, the supply of building land for commercial use has declined and large new areas are not ready for development. Increasing land prices are the result, pushing many companies into the surrounding countryside where sufficient land at acceptable prices is usually available. Sometimes in city-centre districts the land and rental prices are extremely high (see the Frankfurt case study, Ch. 11.2), so that even office-users shift their administrative staff into the surrounding countryside.

By contrast, in economically weak, peripheral and older industrialized regions the supply is sufficient, sometimes even with oversupply causing artificially low land prices. In the latter regions, further market failures happen with regard to contaminated brownfield land. These areas often remain unused, while greenfield sites remain in demand, because even low land prices do not permit economic decontamination (as the comparison of the Dortmund and Düsseldorf case studies shows, Chs 7.3 and 11.1).

REFERENCES

Entries in **bold type** are edited volumes of contributions cited more than once.

Ache, P., F-J. Ingemey, K. R. Kunzmann, H-J. Bremm 1989. *Die regionale Entwicklung süddeutscher Verdichtungsräume – Parallelen zum Ruhrgebiet?* Essen: Kommunalverband Ruhrgebiet (KVR).

Ache, P., H-J. Bremm, K. R. Kunzmann 1990. *The Single European Market – possible impacts on the spatial structure of the Federal Republic of Germany.* IRPUD Arbeitspapier 91. Dortmund: University of Dortmund.

Aengevelt, L. 1990a. *Vergleichende Untersuchung Gewerbeparks – Das Beispiel des Großraums Düsseldorf* (market report). Düsseldorf: L. Aengevelt.

Aengevelt, L. 1990b. *Immobilienmarktbericht Frankfurt/Main, Stand März/April* (market report). Düsseldorf: L. Aengevelt.

Aengevelt, L. 1990c. *Immobilienmarktbericht Köln* (market report, no. 6). Düsseldorf: L. Aengevelt.

Arras, E. A. 1979. *Wohnungspolitik und Stadtentwicklung, Teil 1*. Bonn: Bundes minister für Raumordnung, Bauwesen und Städtebau, Schriftenreihe Städtebauliche Forschung, no. 03.084.

Arras, E. A. 1983. *Wohnungspolitik und Stadtentwicklung, Teil 2*. Bonn: Bundes minister für Raumordnung, Bauwesen und Städtebau, Schriftenreihe Städtebauliche Forschung, no. 03.094.

Arras, H. E. 1981. Städtebau-, Wohnungs- und Bodenpolitik – Brauchen wir ein integriertes Entwicklungskonzept?' *Innere Kolonisation*, 215–17.

Bartholmai, B., M. Melzer, E. Schulz 1991. Aktuelle Tendenzen der Wohnungs marktentwicklung in Deutschland. *Informationen zur Raumentwicklung* 5, 301.

Battis, U. 1987. *Öffentliches Baurecht und Raumordnungsrecht*, 2nd edn. Stuttgart: W. Kohlhammer.

Bauer, M. & H. W. Bonny 1987. *Flächenbedarf von Industrie und Gewerbe – Bedarfsrechnung nach Gifpro*. Dortmund: Institut für Landes- und Stadtentwick-lungsforschung des Landes Nordrhein–Westfalen.

Baur, F. 1989. *Lehrbuch des Sachenrechts*, 15th edn. München: C. H. Beck.

Beuerlein, J. 1989. Nutzung der Bodenfläche in der Bundesrepublik Deutschland – Erste Ergebnisse der Flächenerhebung 1989. *Wirtschaft und Statistik* 6, 389–93.

Bielenberg W. & W. Kleiber 1992. *Eigentum an Grund und Boden in den neuen Ländern*, 1 Auflag. München: Stand February.

Böltken, F. & K. P. Schön 1989. Zur Entwicklung der Struktur von Städten in der

REFERENCES

Bundesrepublik Deutschland. *Informationen zur Raumentwicklung* **11/12**, 823–43.

Brake, K. 1986. Das "Süd-Nord-Gefälle" als Ausdruck epochaler Strukturveränderungen in Produktion und Territorium. *RaumPlanung* **34**, 171–4.

Brockhoff Zadelhoff 1991. *Der Markt für Büro- und Hallenflächen im Ruhrgebiet 1991*. Essen: Brockhoff Zadelhoff.

Bundesamt für Statistik 1991. *Ergebnisse der Volkszählung 1987, Stand: 10.06.91*. Wiesbaden: Budnesamt für Statistik.

BMBau (Bundesminister für Raumordnung, Bauwesen und Städtebau) 1983. *Baulandbericht 1983*. Schriftenreihe Städtebauliche Forschung no. 03.100. Bonn–Bad Godesberg: BMBau.

BMBau 1986. *Baulandbericht 1986*. Schriftenreihe Städtebauliche Forschung no. 03.116. Bonn–Bad Godesberg: BMBau.

BMBau 1989. *Das Programm für eine Million neue Wohnungen*. Bonn–Bad Godesberg: BMBau.

BMBau 1990a. *Raumordnungsbericht 1990*. Bonn–Bad Godesberg: BMBau.

BMBau 1990b. *Haus und Wohnung im Spiegel der Statistik*. Bonn–Bad Godesberg: BMBau.

BMBau 1990c. *Bau- und Wohnfibel*. Bonn–Bad Godesberg: BMBau.

Cihan, A., U. Becker, C. Cremer 1990. Polarisierung der Innenstadt – Tertiär isierungsprozesse und Entwicklungstendenzen in westdeutschen Stadtregionen. *RaumPlanung* **50**, 127–35.

Citron, R. 1991. *Getting into Europe – strategic planning for international tax, raising finance, and performance monitoring*. London: Kogan Page.

Creutz, H. 1987. *Bauen, Wohnen, Mieten – Welche Rolle spielt das Geld?* Hannoversch Munden: Gauke-Fachverlag für Sozialökonomie.

Dangschat, J. S. & J. Blasius (eds) 1990. *Gentrification. Die Aufwertung innenstadtnaher Wohnviertel*. Frankfurt: Campus.

Dehnen, H. 1990. *Eigener Herd ist Goldes wert – Ratgeber zum Bau und Erwerb von Wohneigentum*, 14th edn. Bonn: Domus.

Deutsche Grundbesitz-Investmentgesellschaft mbH 1990. *Grundbesitz-investment: Rechenschaftsbericht 1989/90*. Frankfurt: Deutsche Grundbesitz-Investmentgesellschaft mbH.

Deutsche Immobilien Fonds AG 1990. *DIFA-Grund: Rechenschaftsbericht zum 30.09.1990*. Hamburg: Deutsche Immobilien Fonds AG.

Deutsche Immobilienpartner (DIP) 1990. *Markets and Facts '90* Düsseldorf: Deutsche Immonilien Fonds AG.

Deutscher Städtetag 1990. Fertiggestellte Wohnungen in neu errichteten Wohnungen und Nichtwohnungen 1989 – Ergebnisse der Bautätigkeitsstatistik der statistischen Landesämter. *Der Städtetag* **11**, 821–31.

Deutscher Städtetag. Various years. *Statistische Jahrbüche deutscher Gemeinden*. Köln: Deutschwe Städtetag.

Deutsches Institut für Wirtschaftsforschung 1991. Vierteljahresdaten zur Wohnungs- und Bauwirtschaft. *Bundesbaublatt* **5**, 330–2.

Dierks, L. & H. Kirchner 1989. Der regionale Koordinationsbedarf wächtst – aber wer koordiniert? Infrastrukturplanung im Verdichtungsraum am Beispiel Region Mittlerer Neckar. *Informationen zur Raumentwicklung* **1**, 13–23.

Dieterich, H. 1990a. *Baulandumlegung*, 2nd edn. München: C. H. Beck.

Dieterich, H. 1990b. Kommentar zu 194 BauGB. See Ernst, Zinkahn, Bielenberg (1990), ??–??.

Dieterich, H. 1991a. Bodenordnung und Bodenpolitik. In *Kompendium der Wohnungswirtschaft*, H. W. Jenkis (ed.), 250–75. München: Oldenbourg.

Dieterich, H. 1991b. Neues Bodenrecht? *Kommunale Steuer-Zeitschrift* **7**, 121–6.

Dieterich, H., B. Dieterich-Buchwald, E. Dransfeld, F-J. Lemmen, W. Voß 1991. *Wirkungsforschung zur Baulückenbebauung*. Bonn: Bundesminister für Raumordnung, Bauwesen und Städtebau.

Dieterich, H. & E. Dransfeld 1992. Germany. In *The industrial property markets in Western Europe*, B. Wood & R. H. Williams (eds), 27–65. London: E. & F. N. Spon.

Dieterich, H. & J. Hucke 1985. Funktionsweise des Bodenmarktes im Umland von Großstädten. Berlin/Dortmund: unpublished research project.

Dietz, R. & C. Ernst 1991. Der Büroflächenmarkt im Ruhrgebiet. Diplomarbeit (unpublished dissertation), Fakultät Raumplanung, Universität Dortmund.

Düsseldorf Magazin 1989. Zum ersten Mal – Düsseldorf & Partner. *Düsseldorf Magazin* **4**, 44–45.

Eekhoff, J. 1984. Wohnungsbaufinanzierung unter wechselnden wirtschaftlichen Bedingungen (Teil 1-4). *Der langfristige Kredit* **11**, 334–8; **13**, 415–20; **14**, 436–44; **15**, 474–6.

Epping, G. 1977. *Bodenmarkt und Bodenpolitik in der Bundesrepublik Deutschland*. Berlin: Duncker & Humblot.

Erbguth, W. 1989. *Bauplanungsrecht*. München: C. H. Beck.

Ernst, W. 1984. Bodenpolitik, Bodenrecht und Planungsrecht in Stadt und Land *Vermessungswesen und Raumordnung* 1984, 366–78.

Ernst, W., W. Zinkhahn, W. Bielenberg (eds) 1990. Baugesetzbuch – Kommentar. München: C. H. Beck.

Euler, M. 1991. Grundvermögen privater Haushalte Ende 1988 – Ergebnisse der Einkommens- und Verbraucherstichprobe. *Wirtschaft und Statistik* **4**, 277–84.

EUREGIO Maas-Rhein 1991. *Immobilienerwerb in der Euregio – Eine Information für die Bürger der Euregio Maas-Rhein*. Aachen: Beschwerdestelle Regio Aachen.

Falk, B. (ed.) 1985. *Immobilienhandbuch – Wirtschaft, Recht, Bewertung*. Stuttgart: W. Kohlhammer.

Falk, B. (ed.) 1987. *Gewerbe-Immobilien*. Landsberg/Lech: Moderne Industrie.

Finanzminister des Landes Baden-Württemberg 1990. Verwaltungsvorschrift über die verbilligte Abgabe von landeseigenen Grundstücken vom 29. 10. 1990. *Bundesbaublatt* 1991, 118.

Fleischer, H. 1990. Ausländer 1989. *Wirtschaft und Statistik* **8**, 540–44.

Friauf, K. H., W. K. Risse, K-P. Winters 1978. *Der Beitrag steuerlicher Maß*

REFERENCES

nahmen zur Lösung der Bodenfrage. Bonn–Bad Godesberg: Bundesminister für Raumordnung, Bauwesen und Städtebau, Schriftenreihe "Städtebauliche Forschung" no. 03.064.

GEWOS 1990a. Institut für Stadt-, Regional- und Wohnforschung GmbH. GEWOS-Immobilienmarktanalyse 1990 – Materialien zur Pressekonferenz am 28. 6. 1990 in Hamburg. Hamburg.

GEWOS 1990b. Institut für Stadt- und Regionalplanung GmbH *Immobilienmarkt analyse* 1990. Hamburg: Managementkurzfassung.

Greiner, H. 1991. Die Verflechtung von Wohnen und Arbeiten in der Region. *Magazin Wirtschaft* (Industrie und Handelskammer Region Stuttgart) 1, 17–18.

Günther, A. 1988. Die Planung des Technologiegebietes in Dortmund. *Stadtbauwelt* 99, 1556–64.

GA Stuttgart (Gutachterausschuß für die Ermittlung von Grundstückswerten in Stuttgart) 1980–90. *Der Grundstücksmarkt in Stuttgart.* Stuttgart: Stadtmessungsamt.

GA Hildesheim (Gutachterausschuß für Grundstückswerte für den Bereich des Landkreises Hildesheim) 1985–90. *Berichte über der Grundstücksmarkt im Landkreis Hildesheim.* Hildesheim, Alfeld: GA Hildesheim

GA Frankfurt (Gutachterausschuß für Grundstückswerte) 1990. Der Frankfurter Grundstücksmarkt 1988/89. *Frankfurter Statistische Berichte* 3.

GA München (Gutachterausschuß für Grundstückswerte München) 1990.Jahres bericht über den Grundstücksverkehr und die Preisentwicklung im Stadtgebiet München. München: Stadt München.

Güttler, H. 1987. Die Baulandpolitik steht vor neuen Aufgaben – zum Bauland bericht 1986. *Vermessungswesen und Raumordnung* 1987, 58–64.

Güttler, H. & W. Kleiber 1989. Aktuelle Entwicklungstendenzen auf dem Grundstücksmarkt. *Bundesbaublatt* 5, 236–44.

Haasis, H-A. 1987. *Bodenpreis, Bodenmarkt und Stadtentwicklung.* München: Minerva Publikationen.

Häußermann, H. & W. Siebel 1987. *Neue Urbanität.* Frankfurt: Suhrkamp.

Hecking, G. 1988. *Bevölkerungsentwicklung und Siedlungsflächenexpansion. Entwicklungstrends und Perspektiven am Beispiel der Region Mittlerer Neckar.* Stuttgart: Karl Krämer.

Heinlein, D. 1991. Ergebnisse der laufenden Lohnstatistik für 1990. *Wirtschaft und Statistik* 4, 285–92.

Heinz, W. 1990. *Stadtentwicklung und Strukturwandel. Einschätzungen kommunaler und außerkommunaler Entscheidungsträger.* Stuttgart/Berlin/Köln: H. Kohlhammer.

Hennings, G.& E. von Einem 1988. *Möglichkeiten und Formen der Berücksicht igung und Eingliederung gewerbepolitischer Förderstrategien in Städtebaupolitik und Stadtentwicklung – Fallstudie Dortmund.* Bonn: BMBau.

Hennings, G., T. Friederichs, R. Kalscheuer 1989. *Kaiserlei – Bürostadt am Main – Entwicklung und Vermarktung des Kaiserlei-Gebietes,* vol. I (unpublished survey). Dortmund.

Hennings, G., T. Friederichs, R. Kalscheuer 1990. *Kaiserlei - Bürostadt am Main - Entwicklung und Vermarktung des Kaiserlei- Gebietes*, vol. II (unpublished survey). Dortmund.

Heuer, H. 1985. *Instrumente kommunaler Gewerbepolitik - Ergebnisse empirischer Erhebungen*. Stuttgart: Kohlhammer.

Hintzsche, M. 1990. Grundstückspreise in Stuttgart - unbezahlbar oder Ausdruck eines funktionierenden Grundstücksmarktes? *Allgemeine Vermessungsnachrichten* **7**, 264-76.

Hooper, A. J. 1989. Federal Republic of Germany. In *Planning control in Western Europe* (Department of the Environment), 255-335. London: HMSO.

Hüttenrauch, C. & W. Schaar. Preise für Wohnbauland - Umsätze und Preisentwicklung auf dem Grundstücksmarkt. *Der Städtetag* **10**, 691-700.

IHK 1990. (Industrie- und Handelskammer Region Stuttgart). Stuttgart im Standort-Wettbewerb - Acht Städte und Regionen im Vergleich. Stuttgart: Industrie- und Handelskammer Region Stuttgart.

IRPUD 1989. (Institut für Raumplanung) *Die regionale Entwicklung süddeutscher Verdichtungsräume - Parallelen zum Ruhrgebiet? Vol. 2: Materials, tables, literature*. Essen: Kommunalverband Ruhrgebiet (KVR).

Jones Lang Wootton 1989. *Market report: West Germany 1989, International Version*. Hamburg: Jones Lang Wootton.

Jones Lang Wootton 1990a. *City report: Frankfurt*. Frankfurt: Jones Lang Wootton.

Jones Lang Wootton 1990b. *Marktbericht: Bundesrepublik Deutschland 1990/Market report West Germany 1990*. Hamburg: Jones Lang Wootton.

Kamphausen, M. 1989. Gewerbeparks wachsen auf alten Industriebrachen. *Düsseldorf Magazin* **4**, 14-16.

Kleiber, W., J. Simon, G. Weyers 1991. *Recht und Praxis der Verkehrswertermittlung von Grundstücken*. Köln: Bundesanzeiger.

Kreibich, B., V. Kreibich, G. Reesas 1989. *Gutachten zum Wohnungsgesamtplan Köln - Grundszenario der Wohnungsmarktentwicklung* (unpublished survey). Dortmund.

Kreibich, V. 1990. Die Gefährdung preisgünstigen Wohnraums duch wohnngspolitsche Rahmenbedingungen. See Dangschat & Blasius (1990), 51-69.

Kujath, H. J. 1988. Polarisierung des Lebens in der wirtschaftlich stagnierenden Stadt. Zu den sozialräumlichen Folgen des wirtschaftlichen Strukturwandels. *RaumPlanung* **40**, 35-9.

Kunzmann, K. R. 1984. The Federal Republic of Germany. In *Planning in Europe: urban and regional planniing in the EEC*, R. H. Williams (ed.), 8-25. London: Allen & Unwin.

Kunzmann, K. 1986. Die städtebauliche Erneuerung und Aufwertung von alten Industrie- und Gewerbegebieten (IRPUD working paper). Dortmund: Universität Dortmund.

REFERENCES

Landesamt für Datenverarbeitung und Statistik Nordrhein–Westfalen 1987. *Berufs- und Ausbildungspendler in Nordrhein–Westfalen, Ergebnisse der Volkszählung 1987*. Düsseldorf: Landesamt für Datenverarbeitung und Statistik Nordrhein-Westfalen.

Landeshauptstadt Düsseldorf 1989. *Düsseldorf & Partner – Ihr Standort in Europa*. Düsseldorf: Landeshauptstadt Düsseldorf.

Minister für Landes- und Stadtentwicklung 1984. Verwaltungsvorschrift zur Landesbauordnung 29. 11. 1984 (MBl. NW 1984, 1954).

Model, O., C. Creifelds, G. Lichtenberger 1991. *Staatsbürgertaschenbuch*, 25th edn. München: C. H. Beck.

Müller International 1990. *Büro-Markt-Bericht: Bundesrepublik Deutschland*. Frankfurt: Muller International.

Müller-Kleißler, R. & D. Rach 1989. Der Baulandmarkt in der Bundesrepublik Deutschland. *Informationen zur Raumentwicklung* **6/7**, 401–18.

Neisser, H-J. 1989. Der "Blaumann" auf dem Rückzug. *Düsseldorf Magazin* **4**, 6–8.

OECD 1987. Revenue statistics of OECD member countries 1965–87. Paris: OECD.

Peschke, B. & W. Hacker 1987. *Untersuchung zur flughafenbezogenen Ansiedlungs-nachfrage* (unpublished survey) Frankfurt and Bad Homburg: Umlandverband Frankfurt.

Pfeiffer, U. 1981. *Die Wirkung einer veränderten Besteuerung auf Baulandangebot und rationelle Nutzung aus wohnungswirtschaftlicher Sicht*. Bon-Bad Godesberg: BMBau.

Platz, J. 1989. *Immobilien-Management – Prüfkriterien zu Lage, Substanz, Rendite*. Wiesbaden: Gabler.

Pöschl, H. 1990. Singles – Versuch einer Beschreibung. *Wirtschaft und Statistik* **10**, 703–8.

Preisinger, W. 1991. Ein Weg zur langfristigen Kundenverbindung – Immobilien vermittlung im geschlossenen Verband. *Der langfristige Kredit* **8**, 255–6.

Rach, D. & A. Gütelhöfer 1990. Erste Ergebnisse der Flächenerhebung 1989 – Trendwende im Flächenverbrauch? *Mitteilungen der Bundesforschungsanstalt für Landeskunde und Raumordnung* **5**, 3–4.

Raschke, W-D. 1991. Der gläserne Bauherr – Zur LBS-Wohneigentumsstudie 1990. *Der langfristige Kredit* 8, 257–60.

Research Project "EuProMa" 1991. Aims and criteria for judging land and property markets (unpublished working paper). Dortmund.

RDM (Ring Deutscher Makler) 1991. Mietpreisspiegel 1990. *Informationsdienst und Mitteilungsblatt des VHW* **5**, 51–2.

Ricking H-H. 1992. Verkehrwertgutachten für Immobilien in den Beitrittsländern – Probleme bei magelhaften Unterlagen und fehendem Markt, Loesungsversuche durch Anpassungsfaktoren. *Allgemeine Vermessungsnachtrichten* **5**, 201.

Robert, J. 1983. *Räumliche Planung in Belgien, den Niederlanden und Nordrhein-Westfalen*. Dortmund: Institut für Landes- und Stadtentwicklungsforschung des Landes Nordrhein-Westfalen (ILS).

Rössler, R., I. Langner, I. Simon, W. Kleiber 1990. *Schätzung und Ermittlung von Grundstückswerten*, 6th edn. Neuwied, Frankfurt: Luchterhand.

Runderlaß des Ministeriums für Umwelt, Raumordnung und Landwirtschaft 1990. Abstände zwischen Industrie-bzw. Gewerbegebieten im Rahmen der Bauleitplanung (Abstandserlaß). 21. 03. 1990 (MBl. 1990 S. 504).

Schmidt-Eichstaedt, G. 1987. *Einführung in das neue Städtebaurecht – ein Handbuch*. Stuttgart: W. Kohlhammer.

Schöffel, R. 1991. Einheitswerte der gewerblichen Betriebe 1986. *Wirtschaft und Statistik* **2**, 128–34.

Scholland, R. 1987. Die Bedeutung der Funktionsweise des Bodenmarktes für den Verlauf der regionalen Siedlungsentwicklung. *Allgemeine Vermessungsnachrichten 1987*, 72–83.

Schönhofer, R. 1991. *Haus- und Grundbesitz in Recht und Praxis*. Freiburg: Rudolf Haute.

Schubert, W. 1991. Insel der Seligen. Deutschland avanciert zum besten Standort. *Wirtschaftswoche*, 26 April 1991, 72–86.

Schulz, A. 1990. Die Aufwertung innerstadtnaher Wohnviertel in Köln. See Dangschat & Blasius (1990), 185–96.

Schwenk, W. 1991. Zu den grundlegenden Problemen der Ermittlung von Verkehrwerten in den neuen Bundesländern. *BDVI Forum* **4**, 219.

Schwenk, W. 1992. Alles Rechte, neues Unrecht. Der Spiegel **27** (29), June 1992.

Seele, W. 1992. Bodenorderische Probleme in den neuen Bundesländern. *Vermessungswesen und Raumordnung* **2–3**, 73.

Simon, J. 1991. *Recht und Praxis der Verkehrswertermittlung*. von Grundstücken (eds) Kleiber W, Simon J. and Weyers G Köln: Bundesanzeiger.

Sinz, W. J. 1989. Regionale Wettbewerbsfähigkeit und Europäischer Binnenmarkt. *Raumforschung und Raumordnung* **1**, 10–21.

Sommer, B. & H. Fleischer 1991. Bevölkerungsentwicklung 1989. *Wirtschaft und Statistik* **2**, 81–8.

Stadt Dortmund 1990. *Technologie Park Dortmund, Documentation no. 4*. Dortmund: Stadt Dortmund.

Stadt Frankfurt 1990a. *Statistisches Jahrbuch Frankfurt am Main 1990*. Frankfurt: Stadt Frankfurt.

Stadt Frankfurt 1990b. Frankfurter Monats- und Quartalszahlen. *Frankfurter Statistische Berichte* **3**, 83–92.

Stadt Hildesheim 1979. *Flächennutzungsplan*. Hildesheim: Stadt Hildesheim.

Stadt Hildesheim.1980. Liegenschaftsamt, *Richtlinien zur Vergabe von Eigenheim grundstücken*. Hildesheim: Stadt Hildesheim.

Stadt Hildesheim 1981. *Begründungen zu den Bebauungsplänen nr. 197 A, B, C, D*. Hildesheim: Stadt Hildesheim.

Stadt Hildesheim 1989. Einwohneramt, *Eckdaten zur Volkszählung 1987*. Hildes heim: Stadt Hildesheim.

Stadt Hildesheim 1990. Einwohneramt, *Grunddaten zur Verkehrssituation in Hildesheim*. Hildesheim: Stadt Hildesheim.

Stadt Hildesheim 1991a. Einwohneramt, *Informationen zur Stadtentwicklung*. Hildesheim: Stadt Hildesheim..

Stadt Hildesheim 1991b. *Schriftiche Auskunft Vermessungsam Hildesheim*. Hildes heim: Stadt Hildesheim.

Stadt Köln 1989. *Entwicklungskonzept Innenstadt*. Köln: Stadt Köln.

Stadt Köln 1991. *Informationssystem 1978-1991*. Köln: Stadt Köln.

Stadt Stuttgart 1974. *Flächennutzungsplan 1974*. Stuttgart: Stadt Stuttgart.

Stadt Stuttgart 1980, 1983, 1986. *Begründung zum Bebauungsplan Stammheim-Süd (Sta 70), (südlicher Teilbereich) (Sta 91), Stammheim-Süd II (Sta 95)*. Stuttgart: Stadt Stuttgart.

Stadt Stuttgart 1985. Sonderprogramm preisgünstiges Wohnungseigentum. *Amtsblatt* 39. Stuttgart: Stadt Stuttgart.

Stadt Stuttgart 1990a. Statistisches Amt. Jahresübersicht 1990. *Statistische Blätter* 47 Stuttgart: Stadt Stuttgart.

Stadt Stuttgart 1990b. *Bodenordnungsverfahren – Jahresbericht 1990*. Stuttgart: Stadt Stuttgart.

Stadt Stuttgart 1991. Grundstücksmarkt in Stuttgart. *Amtsblatt* 18. Stuttgart: Stadt Stuttgart.

Statistisches Bundesamt 1990. *Statistisches Jahrbuch 1990 der Bundesrepublik Deutschland*. Stuttgart: Metzel-Poeschel.

Steuergesetze I – Textsammlung, 3 October 1990. München: C. H. Beck.

Strohm, W. 1991. Sozialprodukt im bisherigen Gebiet der Bundesrepublik Deutschland im Jahr 1990. *Wirtschaft und Statistik* 1, 17–27.

Thielges, R. 1991. Zur Verkehrswertermittlung ehemals militärisch genutzter Grundstücke. *Nachrichtenblatt der Vermessungs- und Katasterverwaltung Rheinland Pfalz* 1, 27–30.

Troll, M. 1977. *Privater Hausbesitz im Steuerrecht*. Stuttgart/Wiesbaden: Forkel.

Umlandverband Frankfurt 1989a. Erläuterungsbericht zum Flächennutzungsplan des Umlandverbandes Frankfurt. Frankfurt: Umlandverband Frankfurt.

Umlandverband Frankfurt 1989b. *Flächenbedarf von Arbeitsstätten – Entwicklungstendenzen in ausgewählten Wirtschaftsbereichen* (unpublished survey). Frankfurt.

Verbraucher-Zentrale NRW 1988. *Der Wohnungskauf – Probleme und Lösungen*. Düsseldorf: Verbraucher-Zentrale NRW

VHW 1991. *Informationsdienst und Mitteilungsblatt des Volksheimstättenwerkes (VHW)* 5, 51.

von der Heide, K-H. 1988. Flächenmaßstab und Flächenbeitrag nach 58 Baugesetzbuch – Vergangenheit, Gegenwart, Zukunft. *Vermessungswesen und Raumordnung* 4, 228–34.

von der Heide, K-H. 1989. Entwicklung der Bodenordnung in Stuttgart. *Vermes sungswesen und Raumordnung* **4/5**, 267–74.

Von Dornbusch, H-L. & L. Th. Jasper 1988. *Steuervorteile für Immobilienanleger*, 2nd edn. München: Beck-Rechtsberater im DTV.

Von Dornbusch, H-L., L. Th. Jasper, D. Piltz 1990. *Steuervorteile durch Haus- und Wohnbesitz*, 4th edn. München: Beck-Rechtsberater im DTV.

Weirich, H-A. 1985. *Grundstücksrecht*. München: C. H. Beck.

Weyers, G. 1990. Neue Grundsätze zur Ableitung des Beleihungswertes von Grundstücken bei Kredit- und Versicherungsinstituten. *Grundstücksmarkt und Grundstückswert* **2**, 74–82.

Winter, H. 1991. Mieten in der Bundesrepublik Deutschland – Ergebnisse der Gebäude- und Wohnungszählung von 1987 und ein Vergleich mit der Situation von 1968. *Wirtschaft und Statistik* **3**, 169–75.

Wirtschaftsförderungsamt der Stadt Düsseldorf 1990. *Düsseldorf – internationales Wirtschaftszentrum in Europa*. Düsseldorf: Stadt Düsseldorf.

Würdemann, G. & T. Pütz 1990. Räumliche Aspekte der ungebremsten Motorisierung. *Mitteilungen der Bundesforschungsanstalt für Landeskunde und Raumordnung* **4**, 4–5.

Zeitschrift für das gesamte Kreditwesen 1990. *Die Finanzierungshilfen des Bundes und der Länder an die gewerbliche Wirtschaft einschließlich der Finanzierungshilfen internationaler Institutionen und der DDR-Programme*, vol. 1. Frankfurt: Fritz Knapp.

Zinkahn, W. 1982. Einführung zum Bundesbaugesetz. In BBauG, 15. Edition. München: Beck-Texte im DTV.

Zinkahn, W. & W. Söfker 1990. Einführung zum Baugesetzbuch. In BauGB, 20. Edition. München: Beck-Texte im DTV.

LEGAL INSTRUMENTS

Baugesetzbuch 8. 12. 1986 (BGBl. **I**, 2253) – (planning code/town and country planning act) – changed by reunification treaty from 31. 8. 1990 (BGBl. **II**, 889).

Baunutzungsverordnung (BauNVO) – (land-use ordinance) – 19. 07. 1989 (BGBl. **I**, 132), changed by reunification treaty from 31. 08. 1990, (BGBl. **II**, 889, 1122).

Bauordnung für das Land Nordrhein–Westfalen – Landesbauordnung (BauONW) – (building code) – 26. 06. 1984, (GVBl. NW, 419), changed 20. 06. 1989, (GVBl. NW, 432).

Bauzulassungsverordnung (BauZVO) – (ordinance for the permission of buildings in the former GDR) – 20. 06. 1990 (GBl. DDR **I**, 739), changed 20. 7. 1990 (GBl. DDR **I**, 950).

II. Berechnungsverordnung 20. 08. 1990 (BGBl. **I**, 1813).

Bewertunggesetz (BewG) 01. 01. 1991 (BGBl. **I**. 230) – (act on valuing properties).

Bundesimmissionsschutzgesetz (BImSchG) – (federal act for protection against pollution) – 14. 05. 1990, (BGBl. **I**, 880), changed by reunification treaty from 31. 08. 1990 (BGBl. **II**, 885).

Bürgerliches Gesetz (BGB) – (civil code) – 18. 08. 1896, (RGBl., 195, BGBl. **III**, 400–402).

Gebührenordnung für die Vermessungs- und Katasterbehörden in Nordrhein–Westfalen (VermGebO) 26. 4. 1973 (GVBl. NW., 308) – (scale for changes for surveying and cadaster), changed 18. 2. 1990 (GVBl. NW., 58).

Gesetz über das Wohnungseigentum und das Dauerwohnrecht (Wohnungseigentumsgesetz) 15. 03. 1951 (BGBl. **I**, 175) – (act for condominium law) – changed 14. 12. 1984 (BGBl. **I**, 1493).

Gesetz über Kapitalanlagegesellschaften (KAGG) 14. 1. 1970 (BGBl. **I**, 127), changed 1. 8. 1990 (BGBl. **II**, 889, 976).

Gesetz zur Regelung der Miethöhe (Miethöhengesetz) 18. 12. 1974 (BGBl. **I**, S. 3604), changed 20. 12. 1982 (BGBl. **I**, S. 1912) – (act for rent control)

Gesetz zur Regelung der Wohnvermittlung vom 4. 11. 1971 (BGBl. **I**, 1745) – (act to regulate the placing of flats), changed 17. 12. 1990 (BGBl. **I**, 2840).

Gewerbeordnung 01. 01. 1987 (BGBl. **I**, 425) – (act governing trade and industry), changed 28. 06. 1990 (BGBl. **I**, 1221).

Grundbuchordnung (GBO) – (land register act) – 05. 08. 1935 (RGBl. **I**, 1073, BGBl. **III**-3 Nr 315-11).

Honorarordnung für Architekten- und Ingineurleistungen (HOAI) 17. 9. 1976 (BGBl. **I**, 2805) – (scale of charges for architects), changed 17. 2. 1988 (BGBl. **I**, 1359).

Hypothekenbankgesetz 19. 12. 1990 (BGBl. **I**, 2898).

Preisangabeverordnung 14. 3. 1985 (BGBl. **I**, 580).

Raumordnungsgesetz 19. 7. 1989 (BGBl. **I**, 1461) – (federal act for spatial policy) – changed by reunification treaty from 31. 8. 1990 (BGBl. **II**, 889).

Verbot der Zweckentfremdung von Wohnraum. Art. 6 des Gesetzes zur Verbes serung des Mietrechts und zur Begrenzung des Mietanstiegs 04. 11. 1971 (BGBl. I, S. 1745). – (regulation to prevent the change of use of living space)

Verordnung über das Erbbaurecht 15. 1. 1919 (RGBl., 72) – (decree on the hereditary long leasehold) – changed 08. 06. 1988 (BGBl. I, 710).

Verordnung über das Erbbaurecht (ErbbRVO) – (ordinance for the hereditary leasehold) – 15. 01. 1919, (RGBl. 72, 122, BGBl. III4 Nr 403–6), changed 8. 6. 1988 (BGBl. I, 710).

Verordnung über Grundsätze für die Ermittlung der Verkehrswerte von Grund stücken (Wertermittlungs-Verordnung – WertV) 06. 12. 1988 (BGBl. I, 2209) – (federal decree on valuation of land and property).

Verordnung über die Pflichten der Markler, Darlehens- und Anlagenvermittler, Bauträger und Baubetreuer (Makler- und Bauträgerverordnung – MaBV) 07. 11. 1990 (BGBl. I, 2479) – (federal decree on real estate agents and building construc tors).

Versicherungsaufsichtsgesetz (VAG) 13. 10. 1983 (BGBl. I, 1261) – (federal act for controlling insurance companies).

Wohnungsbauerleichterungsgesetz (WoBauErlG) – (act on facilitating housing) – 17. 05. 1990, (BGBl. I, 926).

II. Wohnungsbaugesetz 14. 08. 1990 (BGBl. I, 1730).

INDEX OF ENGLISH TERMS

amortization 82, 83, 99–101
architects 31, 85, 91, 111, 112, 115, 148, 172, 183, 198, 209, 218, 223, 227, 232, 233, 234, 242, 266

Baden-Württemberg 17, 21, 54, 97, 110, 121, 149, 151, 259
banks 6, 7, 50, 73, 76–9, 80, 82, 84, 108, 112, 115, 116, 135, 148, 180, 183, 187, 197, 214, 218, 221, 225–7, 230, 235, 240;
see also Sparkassen
Bayern 17, 18, 53, 54, 121
Berlin 2, 17, 21, 27, 28, 30, 35, 50, 51, 53, 54, 75, 82, 93, 103, 129, 200, 254, 259, 260
betterment 41, 42, 66, 67, 72, 73, 128, 141, 166, 168, 193
BfLR 19, 36
Bochum 175, 205
Bonn 74, 205, 236, 257–60
Brandenburg 50, 53, 54
Braunschweig 129
Bremen 2, 17, 27, 131
building societies 76, 78–80, 102, 103, 104
investment 102; *see also* Bausparkassen

cadastre 4, 52, 73, 85; *see also* Kataster
census 74, 216, 238
Certificate of Completion 243
change of use 191, 192, 267
city states 2, 21; *see also* Stadtstaaten
compensation 46–9, 63, 67, 68, 70, 144, 192, 230, 231
compulsory purchase 70, 72; *see also* expropriation
condominiums 5, 6, 69, 100, 102, 103, 140, 146, 147, 149, 202, 203, 239, 266
conservation 4, 191, 194, 231; *see also* Denkmalschutz
contaminated land, *see* derelict land

contractors 64, 71, 79, 81, 110–15, 137, 138, 144–7, 148, 165, 169, 171–3, 187, 197
conversion 7, 32, 44, 45, 61, 69, 74, 128, 191, 196, 197, 201, 216, 227, 235, 236, 240, 241, 243–6; *see also* Umwandlung
co-operatives 47, 50, 82, 111, 117; *see also* Wohnungsbaugenossenschaften
corporation tax 86, 92
counties 3, 4, 62, 74, 129, 191, 236; *see also* Kreis

DDR (former German Democratic Republic) 15, 17, 34, 38, 46–50, 52, 53, 117, 265, 266
former inner German border 140
demolition 191, 192, 208, 211
derelict land 43, 96, 102, 118, 175–7, 204
development fees 85, 95, 133, 144, 166, 168–70, 182
Dortmund 29, 60, 66, 108, 112, 116, 119, 122, 173–7, 179, 180, 182–4, 187, 188, 205, 208, 255, 257, 259–63
Düsseldorf 27, 28, 30, 61, 79, 82, 108, 112–14, 118, 173, 197, 204–208, 210, 212, 213, 217, 236, 254, 255, 257–9, 261, 262, 264, 265

encumbrance 75; *see also* Grundschuld
energy saving 194
environmental assessment 62, 63
EC directive on 63
Essen 175, 205, 257, 258, 261
estate agents 74, 84, 91, 112–15, 128, 197, 198, 211, 232, 238, 240, 267
expropriation 46, 48–50, 70

Frankfurt (Main) 17, 28, 29, 30, 65, 82, 93, 108, 113, 114, 115, 117–18, 128, 152, 155, 176, 191, 205, 213–35, 254,

255
freehold 5, 74

GDP (gross domestic product) 8, 11, 12
gentrification 216, 227, 230, 235, 258
German Democratic Republic, *see* DDR
GFZ (floor-space ratio) 139–41, 144, 145,
165–8, 220,
GNP (gross national product) 8, 47, 75
GRZ (plot ratio) 139–41, 165–6

Hamburg 2, 17, 21, 27, 35, 82, 129, 131,
158, 197, 200, 254, 258, 260, 261
Hannover 27, 129, 131, 133–5
Hessen 17, 18, 54, 214, 216
Hildesheim 66, 81, 108, 111, 112, 115,
116, 129, 131, 133–5, 137–9, 141,
145–7, 148, 260, 263, 264
HOAI (architects' scale charges) 85
housing associations 20, 42, 81, 110, 112,
116, 122, 134, 135, 138, 140, 145–7,
160, 197, 198

income tax 86, 90–93, 97, 100, 101; *see
also* Einkommensteuer
infrastructure 51, 52, 57, 58, 60, 70–72,
94, 95, 102, 115, 154, 176, 187, 204,
206, 213, 218
inheritance 5, 7, 8, 66, 74, 75, 78, 79,
154, 173
    tax 80
inner cities 17, 20, 26, 29, 34, 39, 40,
42, 43, 51, 66, 67, 71, 73, 155, 158,
179, 194, 204, 214, 216, 217, 223, 234,
235, 236, 238, 240, 242, 245, 252, 253
insurance companies 50, 76, 79, 80,
82–4, 92, 99, 112, 113, 135, 138, 144,
146, 147, 176, 218, 219, 240, 267
interest rates 9, 27, 76–78, 80, 97, 117,
147, 214, 253
investment 7, 11, 12, 47, 49–51, 61, 75,
79, 81–83, 90, 95, 97, 99, 100, 103, 107,
111–113, 128, 135, 138, 145, 146, 147,
149, 173, 177, 182, 183, 193, 195, 208,
218–220, 223, 232, 233, 234, 241, 258

Jahn, A. 233

Köln 6, 29, 112, 128, 192, 197, 198,

205, 235–40, 242, 244–6, 257, 258, 260,
261, 263, 264

land assembly 66, 108, 171
land banking 70, 83, 97, 122, 171, 187
land charge 75; *see also* Grundschuld
land register 48, 52, 65, 69, 73, 75, 78,
80, 84, 91, 243, 266; *see also* Grundbuch
land transfer tax 84, 86, 90, 91, 94, 99
land-use plan 58–9; *see also*
Flächennutzungsplan
land values 32, 41, 72, 74, 85, 105–107,
110, 128, 141, 166, 168, 182, 224, 234
landscape plans 4, 44, 151, 182
lease 7, 113, 187, 210, 212, 213
leasehold 5, 74, 108, 110, 138, 161, 170,
173, 174, 267; *see also* Erbbauracht
leasing 6, 7, 91, 93, 174; *see also*
Immobilienleasing
listed buildings 192, 194, 195; *see also*
Denkmalschutz
loan capital 75, 76, 79, 81, 82
local business tax 86, 87, 92, 93; *see also*
Gewerbesteuer
local plans 3, 59, 60, 62, 64, 65, 66–68,
71, 107, 118, 119, 137, 138–41, 144,
145, 147, 153, 161, 165, 168, 183, 185,
186, 193, 223, 225, 227, 230, 231, 234,
235; *see also* Bebauungsplan
London 206, 222, 258, 259, 261

Madrid 30
Mecklenburg-Vorpommern 50, 53
migration 17, 18, 36, 39, 43, 50, 149,
151, 235
minerals 4
Mittlerer Neckar 149, 151, 259, 260
mobility 18–20, 59, 217
mortgages 7, 24, 25, 28, 46, 65, 73,
75–8, 80, 84, 103
München 17, 26–8, 35, 82, 93, 117, 118,
122, 152, 153, 155, 176, 177, 200, 254,
257, 259, 260, 262, 264, 265
municipalities 3–5, 18, 19, 32, 35, 39,
44, 47, 53, 57–60, 62, 64–73, 78, 81, 83,
87, 91, 93–7, 101–103, 104, 105, 107,
108–110, 115, 118, 122, 129, 131, 134,
135, 138, 141, 144, 147, 151, 154, 156,
158, 159, 161, 166, 168–71, 173, 174,

176, 177, 182, 183, 186-8, 191-3, 205, 207, 208, 210, 212, 213, 238, 240, 253, 254; *see also* Gemeinde

Neue Heimat 42, 171
Niedersachsen 18, 53, 110, 129, 131, 137
Nordrhein-Westfalen 3, 39, 40, 44, 63, 77, 96, 101, 180, 182, 183, 187, 197, 205, 236, 257, 262, 263, 266
north/south divide 38, 123
notaries 65, 74, 84, 91, 92, 99

owner-occupation 6, 20, 21, 34, 44, 45, 50, 78, 79, 80, 83, 91, 98, 99, 102, 103, 104, 111, 112, 173, 187, 194, 240, 244
ownership 4-8, 20, 21, 23-5, 39, 40, 43, 44, 46, 48-50, 56, 65-7, 75, 79, 84-6, 90-92, 94, 96-8, 107, 108, 110-12, 114, 117, 138, 139, 141, 144-8, 154, 159, 161, 163, 164, 166, 173, 174, 199, 200, 238-40, 246, 252-4

participation 4, 43, 61, 64, 165, 230
pension funds 82, 83, 92, 225, 227, 233
planning permission 56, 61, 62, 69, 85, 111, 177, 179, 191, 204, 210, 212, 230, 234; *see also* Baugenehmigungsverfahren
population 5, 15, 17, 18, 35, 36, 39, 40, 42, 74, 80, 95, 117, 121, 123, 129, 131, 151, 152, 175, 176, 192, 194, 216, 217, 236
pre-emption 69, 70
preservation 192, 241; *see also* Denkmalschutz
prices 5-7, 8, 11, 25-7, 29, 31, 38-40, 41, 42, 44, 51-3, 56, 69, 70, 71, 73, 74, 75, 84, 87, 90, 92, 93, 94-7, 100, 105, 107, 108, 114, 116, 117, 118, 121, 123, 125, 126, 128, 131, 133, 137, 144-6, 147, 148, 149, 152, 154, 155, 156-9, 161, 169-74, 176-80, 182, 183, 187, 196, 199-203, 208, 219, 220, 222, 225, 226, 227, 233-5, 236, 239, 240, 243, 245, 246, 252-5
property tax 41, 85, 87, 92, 99

redevelopment 40, 42, 43, 44, 96, 101, 102, 177, 192, 193, 195-7, 213, 223, 225, 226, 227, 230; *see also* Sanierung

regional planning 57, 108, 129, 176, 186, 188, 216, 253; *see also* Regionalplan
rents 4-7, 17, 21, 28-31, 40, 42, 43, 45, 46, 48, 50, 78, 79, 81, 84, 97, 101, 102, 103, 104, 107, 111, 113, 114, 116, 134, 145, 146, 147, 148, 155, 156, 170, 171, 173, 174, 180, 187, 191, 194, 197, 200, 201, 209, 210, 216, 218-21, 227, 230, 231, 233, 236, 238-41, 239, 243, 244, 245, 253, 254, 266
replotting 56, 66, 67, 72, 94, 107, 137, 138, 141, 144, 145, 159, 161, 165, 166, 168, 170, 173, 183, 192, 193; *see also* Umlegung
restitution 47-53
Rheinland-Pfalz 17, 18
Rhein-Main agglomeration 2, 200, 214-16, 220; *see also* Umlandverband Frankfurt
Ruhrgebiet 30, 118, 129, 174-6, 188, 205, 236, 257-9, 261

Saarland 17, 21
Sachsen 50, 53
Sachsen-Anhalt 53
Sachsen-Anhalt 50, 53
Schleswig-Holstein 53
SEM (Single European Market) 8, 38, 200, 238
servitudes 65, 73
social housing 29, 44, 70, 104, 148, 158, 191, 216, 238, 239, 254
social values 19, 40
speculation tax 86, 91, 92, 128, 246
Speer, Albert 223, 225, 226
state(s), *see* Land/Länder
Stuttgart 8, 17, 26, 27, 32, 66, 67, 71, 81, 93, 108, 111, 115-18, 122, 149, 151-3, 154-66, 168-71, 197, 200, 253, 254, 257, 259-61, 263, 264
subsidies 42-4, 57, 78, 79, 83, 85, 94-7, 101, 102, 104, 135, 147, 148, 158, 160, 171, 173, 174, 176, 177, 187, 193, 195, 197, 208, 254
housing, *see* Bausparprämie
suburbanization 39, 40, 42, 117, 126, 151, 152, 158, 204

tenancies 6, 7, 192, 238, 243, 244; *see*

also Mietvertrag
tenure 4, 7, 31
Thuringen 50
traffic calming 193; *see also*
 Verkehrsberuhigung
Treuhandanstalt 48

urban renewal 42, 45, 59, 72, 78, 101,
 102, 241–3, 245; *see also* Sanierung

valuation 26, 51, 52, 70, 73, 74, 77, 87,
 105, 107, 139, 166, 177, 193, 196, 220,
 233, 245, 267
Valuation Committee 26, 70, 77, 177,
 220, 233, 245; *see also* Gutachteraus
 schüsse für Grundstuckswerte
value added tax 93–4

wealth tax 85, 90
Wiesbaden 27, 258, 262, 264
working hours 13

yields 79, 81, 82, 219

INDEX OF GERMAN TERMS

Abbruchgebot 193
Ablöseverträge 144, 168, 169
Abstandserlasse 63
Abwägung 63, 64
Allgemeine Immobilien Zeitung 114
Allgemeine Vorkaufsrecht 69
Amtsgericht 84
Anerbenrecht 7
Art der baulichen Nutzung 107
Aufbaugesetz 41
Aufstellungsbeschluß 64, 67
Außenbereich 61, 63
Außenwanderung 17
Ausgleichsbeträge 42
Aussiedler 17

Bauantrag 62
Bauerwartungsland 105, 118, 128, 253
Baufreiheit 56
Baugebot 71, 72, 193
Baugenehmigung 62
Baugenehmigungsbehörde 62
Baugesetzbuch 3, 42, 43, 47, 59, 64, 66, 69, 70, 166, 190, 192, 259, 265, 266
Bauherr 111, 113, 115, 262
Baukindergeld 100, 101
Bauland 105, 107, 118, 121, 177, 260
Baulandinformationssysteme 73
Baulandsteuer 41
Baulast 65
Bauleitplanung 56, 58, 59, 63, 263
Baulinie 182
Baulücken 70
Baunebenkosten 85
Baunutzungsverordnung 59, 60, 190, 266
Bauordnung 62, 225, 266
baureifes Land 119, 121, 123, 128, 253
Bausparen 80, 102
Bausparkassen 76
Bausparprämie 102
Bauvoranfrage 62, 210
Bauwich 65

Bebauungsplan 3, 47, 59–62, 66, 67, 69–71, 110, 128, 180, 183, 186, 188, 204, 212, 223, 264
Beleihungswert 77, 78
Besondere Vorkaufsrecht 69
Bestandsschutz 190
Bewertungsgesetz 87
Binnenwanderung 18
Bodenrichtwerte 52, 107
Bodenschutzprogramm 43
Bodenvorratspolitik 83
Bruttowertschöpfung 11
Bund 2, 3, 78, 79, 87, 91, 95–7, 101–104, 108, 116, 174, 193
Bundesbaugesetz 41, 166, 265
Bundesimmissionsschutzgesetz 63, 161, 190
Bundesnaturschutzgesetz 63
Bundesraumordnungsregionen 74
Bürgerliches Gesetzbuch 4, 65, 190

Dauerwohnrecht 65, 266
Denkmalschutz 191
Deutsche Stadtentwicklungsgesellschaft 110, 115, 197
Dienstbarkeiten 65, 187
Dienstleistungen 11
Diskontsatz 76
Dorferneuerung 43

Einheitswert 24, 25, 87, 92
Einkommensteuer 86
Entwicklungskonzept Hauptverkehrsnetz 242
Entwicklungskonzept Innenstadt 240, 242, 264
Entwicklungsstufen 105
Erbbaurecht 5, 74, 110, 174, 267
Erbschaft-/Schenkungsteuer 86
Erschließungsbeitrag 72
Erschließungsbeitragsfreies Bauland 107
Erschließungsbeitragssatzung 71

Erschließungskosten 168
Erschließungsvertrag 71, 115
Ertragswertverfahren 107
Erwerber 111
Existenzgründer 95

Finanzamt 91
Flächenbeitrag 166, 168, 264
Flächennützungsplan 47, 59, 61, 64, 69,
73, 118, 134, 160, 177, 179, 188, 206,
223, 238, 263, 264
Flächenumlegung 166, 168
Flurkarten 73
Förderweg 103, 104, 170
freie Berufe 91

gemeinnützige Wohnungsbaugesellschaften
112
gemeinnützige Wohnungsunternehmen 81
Gemengelagen 190
Gewerbeaufsicht 4
Gewerbekapitalsteuer 92
Gewerbeparks 83, 114, 257, 261
Gewerbesteuer 86, 87, 92, 254
Grenzregelung 67, 72
Grundbuch 65, 73, 75, 84, 91, 243
Grundbuchordnung 73, 266
Grunderwerbsteuer 86
Grundgesetz 2, 4
Grundschuld 75
Grundsteuer 85
Grundstücksfonds 96, 197
Grundstücksgleiche Rechte 5
Grundvermögen 20, 259
Gutacherausschüsse für Grundstückswerte
74

Hebesatz 87, 93
Hypothekarkredit 75, 76, 79

Immobilienfonds 81, 82
Immobilienleasing 7
Industrieland 119
Innenbereich 60
Innenentwicklung 32, 43

Kataster 73
Kerngebiet 223
Kommunalabgabengesetz 71, 193

Kommunale Selbstverwaltung 3
Kommunale Wirtschaftsförderung 95
konkurrierende Gesetzgebung 2
Körperschaftsteuer 86, 92
kosten- und flächensparendes Bauen 170
Kostenordnung 84
Kreis 3, 4, 62, 74, 129, 191, 236
Kreisfreie Städte 3, 62

Land, Länder 2-8, 12, 14, 15, 17-20, 21,
23, 25-27, 32, 34-6, 38-53, 41, 42, 44,
45-54, 55-67, 69-76, 78, 79, 80, 81,
83-7, 90-100, 101-104, 105-14, 115,
116-19, 121-3, 125, 126, 128, 129, 131,
133-5, 137, 138, 139-41, 144, 145, 149,
151-61, 165, 166, 168-77, 179, 180,
182, 183, 186, 187, 188, 190, 191, 193,
194, 196, 197, 199, 204-206, 208, 210,
212-14, 216, 217, 218-220, 222, 224,
225, 227, 230, 231, 232, 233-6, 238,
240, 241, 242, 243, 252- 5, 259, 262,
264, 265, 266, 267
Landesbausparkasse 79
Landesentwicklungsgeselleschaften 110,
115, 172, 197
Landesimmissionsschutzgesetze 63
Landeswassergesetze 63
Landeswohnungsbauprogramm 170
Landschaftsgesetze 63
Landwirtschaftanpassungsgesetz 49
Lastenzuschuß 78, 102, 104
Liegenschaftsamt 112, 263
Lombardsatz 76

Maß der baulichen Nutzung 107
Mehrwertsteuer 86
Mietervereine 197
Mietspiegel 197
Mietvertrag 6, 7
Mikrozensus 74
Ministerpräsident 2
Modernisierungs- und Instandsetzungsgebot
193

Nacherbe 8
Naturschutzgebiete 63
Neue Heimat 42, 171

Offene Immobilienfonds 81, 82

örtliche Verkehrs- und Grünflächen 166

Pflanzgebot 193
Planungswertausgleich 41

Raumordnungsregionen 20, 36
Raumordnungsverfahren 57
Realteilung 8, 154, 159, 161, 166, 173
Regierungsbezirk 3, 206, 214, 216
Regionalplanung 57–9, 176, 260
Residualverfahren 107
Richtwert 182
Ring Deutscher Makler 74, 114, 262
Rohbauland 107, 119, 123, 253

Sachwertverfahren 107
Sanierung 72, 192, 196
Satzungen 3
SO-Gebiet 186
Sonderausgabenabzug 102
Sondergebiet 186
Sozialpflichtigkeit des Eigentums 68
Sparkassen 76
Spekulationsteuer 86, 128
Städtebauförderung 101
Stadtstaaten 2, 17, 21
steuerbegünstigter Wohnungsbau 79
straßenlandbeitragsfreie Zuteilung 168

Teilungsgenehmigung 69

Umlegung 66, 67, 69, 72, 107, 159, 161
Umwandlung 236, 241
Umweltverträglichkeitsgesetz 63
Unbedenklichkeitsbescheinigung 91
unbeplante Innenbereiche 60
Unterbezirk 236

Veränderungssperre 64, 66–9
Verband Deutscher Makler 114
Verfügungs- und Veränderungssperre 66
Vergleichsmiete 239
Vergleichswertverfahren 107
Verkehrswert 69, 70
Vermogensgesetz 49
Vermögensteuer 85
Verteilungsmasse 66
Volkszählung 74, 258, 262, 264
Vorhaben-und-Erschließungsplan 47

Wasserhaushaltsgesetz 63
Wegerecht 65
Wertermittlungs-Verordnung 105, 107, 267
Wertumlegung 141
Wirtschaftsförderungsamt 112, 174, 206, 265
Wirtschaftsförderungsgesellschaften 174
Wohngeld 104
Wohnumfeldverbesserung 193, 196, 241
Wohnungseigentum 5, 6, 264, 266

Zeitschrift für das gesamte Kreditwesen 95, 265
Zonenrandgebiet 129
Zurückstellung von Baugesuchen 64, 68
Zweckentfremdungsverbot 191, 196
Zwischenerwerber 128, 183